CONSCIOUSNESS IN PHILOSOPHY AND COGNITIVE NEUROSCIENCE

CONSCIOUSNESS IN PHILOSOPHY AND COGNITIVE NEUROSCIENCE

Edited by

Antti Revonsuo
University of Turku

Matti Kamppinen
University of Turku

LEA LAWRENCE ERLBAUM ASSOCIATES, PUBLISHERS
1994 Hillsdale, New Jersey Hove, UK

Lawrence Erlbaum Associates, Inc., Publishers
365 Broadway
Hillsdale, New Jersey 07642

Cover design by Mairav Salomon-Dekel

Library of Congress Cataloging-in-Publication Data

Consciousness in philosophy and cognitive neuroscience / edited by Antti Revonsuo, Matti Kamppinen
p. cm.
"Based on the International Symposium of Consciousness held at the University of Turku...1992"—Pref.
Includes bibliographical references and index.
ISBN 0-8058-1509-0 (alk. paper)
1. Consciousness—Congresses. 2. Philosophy of mind—Congresses. 3. Cognitive neuroscience—Congresses. I. Revonsuo, Antti. II. Kamppinen, Matti, 1961– .
BF311.C654 1994
126—dc20 93-50648
CIP

Books published by Lawrence Erlbaum Associates are printed on acid-free paper, and their bindings are chosen for strength and durability.

Printed in the United States of America
10 9 8 7 6 5 4 3 2 1

Contents

Contributors

Bernard J. Baars, The Wright Institute, Berkeley

Patricia Smith Churchland, Department of Philosophy, University of California at San Diego

Daniel C. Dennett, Department of Philosophy, Tufts University

John Haugeland, Department of Philosophy, University of Pittsburgh

Matti Kamppinen, Department of Cultural Studies, University of Turku, Finland

James Newman, Denver, Colorado

Ilkka Niiniluoto, Department of Philosophy, University of Helsinki, Finland

Vilayanur S. Ramachandran, Department of Psychology, University of California at San Diego

Antti Revonsuo, Department of Philosophy and Center for Cognitive Neuroscience, University of Turku, Finland

Seppo Sajama, Department of Philosophy, University of Joensuu, Finland

John R. Searle, Department of Philosophy, University of California, Berkeley

Raimo Tuomela, Department of Philosophy, University of Helsinki, Finland

Andrew W. Young, MRC Applied Psychology Unit, Cambridge, England

Preface

This book is based on the International Symposium of Consciousness held at the University of Turku. The meeting took place during three warm summer days June 4–6, 1992. The purpose of the symposium was to gather together some of the leading philosophers and cognitive scientists who had recently participated in the discussion of the status of consciousness in science. The question addressed at the symposium was: Is it possible to incorporate consciousness into science? Philosophers have suggested different alternatives: that consciousness should be eliminated from science altogether because it is not a real phenomenon; that consciousness is a real, higher level physical or neurobiological phenomenon; and that consciousness is fundamentally mysterious and beyond human science. At the same time, however, several models or theories of consciousness or the place of conscious processing in the brain have been developed in the more empirical cognitive sciences. It has been suggested that preconscious or unconscious processes must be sharply separated from conscious ones and that this distinction is manifested, for example, in studies of automatic and controlled processing and in certain neuropsychological patients. Models of the functional role of consciousness have been based on these empirical findings. The program of the symposium purported to encourage a dialogue between the philosophical and the empirical points of view. How do different philosophical convictions relate to empirical findings? What kind of implicit, philosophical assumptions guide the modeling of consciousness? What are the plausible alternative solutions to the brain–consciousness problem and how do they enable the unification of biological and psychological sciences?

Do the empirical results support, for example, the elimination of consciousness from science, or is there some converging evidence and a possibility of creating a unified theory of conscious phenomena?

The atmosphere of excitement and the spirit of interaction between conceptual and empirical issues on consciousness in this meeting was something so stimulating that we have also attempted to capture it in this book, and so make it accessible to a wider audience. Thus, the structure of the book is loosely based on the structure of the sessions at the meeting. The purpose of the introductory sections is to familiarize the uninitiated readers with the backgrounds of current debates on consciousness, and the chapters themselves provide a thorough review of theoretical and empirical studies of consciousness at present.

We are indebted to many people and institutions for their help and support during this project. We gratefully acknowledge the funding we have received from the Academy of Finland (contracts #1071132 and #10343) and from the Turku University Foundation for arranging the International Symposium on Consciousness. We owe special thanks to the invited speakers of the symposium who without hesitation accepted our invitation to come to Finland and made the meeting worthwhile. Furthermore, they kindly agreed to contribute to this book; without their favorable cooperation this volume would never have gotten off the ground. We also greatly appreciate those who helped us in arranging the scientific program of the symposium and acted as chairpersons there: Matti Laine, Ilkka Niiniluoto, and Risto Hilpinen. In addition, we thank Kirsti Lagerspetz and Martin Panelius for their kind cooperation with us during the planning stages of the symposium. We warmly thank Maria Vasenkari for preparing the camera-ready version of the book, and, last but not least, Tuija Hovi and Ari Maunu, who kept all the practical strings in their hands during the symposium.

During the preparation and editing of this book, Antti Revonsuo was acting as a Research Associate of the Academy of Finland, which he gratefully acknowledges. Furthermore, he cordially wishes to thank Eira Revonsuo for checking the language of the parts AR contributed in this book, and the friends and colleagues at the Department of Philosophy and the Research Centre for Cognitive Neuroscience, University of Turku, for their friendly support. Matti Kamppinen wants to thank the Department of Cultural Studies, University of Turku, for providing, once again, the resources for realizing another multidisciplinary project.

Antti Revonsuo
Matti Kamppinen

General Introduction: The Riddle of Consciousness

Antti Revonsuo
Matti Kamppinen
University of Turku, Finland

Seppo Sajama
University of Joensuu, Finland

WHAT IS CONSCIOUSNESS?

In a very profound sense, it is consciousness that makes the difference between life and death or existence and nonexistence. What does your whole life consist of but an extended progression of lived conscious experiences? Your stream of awareness started to flow sometime between the womb and the cradle, perhaps with more or less chaotic sensations filling your subjective world. Since then, that world has evolved a lot—now you are surrounded by a relatively coherent perceptual world consisting of colorful, well-formed objects, animals, and people; sounds located in this three-dimensional space, smells in the air and tastes in your mouth, sensations of touch, warmth, cold, pain, and pleasure on your skin and in your body. In your mind, your inner voice almost endlessly comments on your perceptions, intentions, and problems. The stream of thought flows across the perceptual world: "Aha, it looks like rain . . . they didn't predict it, though. All right, let's move . . . do I have everything with me now . . . keys, yes, wallet . . in the bag . . . oh yes, the umbrella, I almost forgot. . . ." Every now and then your awareness becomes immersed in a daydream that might be accompanied by intense visual and auditory imagery, thus causing you to lose your sense of the actual surroundings.

At each moment you have a strong sense of the present, that is the only thing directly accessible to consciousness. The sorrows and joys of yesteryear are long gone from the current experienced reality—at best they can be re-experienced through the feeble memory traces imprinted and buried somewhere inside your brain. Nevertheless, most of the conscious experiences once enjoyed by you have just vanished into thin air and are forever out of your reach. And even a memory, when it is experienced, is a present conscious state. Also any potential future experiences are at the present moment beyond your grasp; all that you can have right now are intentions or dreams concerning possible upcoming circumstances.

Conscious experience, it appears, is really all that matters for your existence. Let us imagine that, since the day you were born, an incurable disease has infected your brain and put it into deep coma for the rest of your life. Consequently, there never was even the slightest sensation of pain or pleasure for you, not the remotest glow of light in the perpetual gloom. The rest of your body was fine, so it developed and grew almost normally, completely oblivious to the fact that it was merely a vacant body devoid of mental phenomena. In this thought experiment, your body does live a life—your pancreas never even noticed the difference, and your spleen could not care less—but *you* are as good as never having existed at all.

So, your body could well have existed without your mind (i.e., without you) ever having existed. But could the body have been behaving like your actual body, and still have been just a nonconscious puppet? Imagine that the clever doctors, after having spotted your irreversible condition, implanted a remote-control device into your brain, thus guiding your body through appropriate behavior with the help of giant computers and a team of scientists, so that no one could tell the difference from the outside. Consequently, this fake-you went through all the actions that you in fact have gone through in your life, and no one ever suspected an ingenious hoax. Everybody took it for granted that somewhere behind the uttered words and bright eyes, there was a subject who could feel his existence in many ways.[1] But they were all wrong, so wrong—you never existed and neither you nor anybody else ever enjoyed even a single conscious experience that in the actual reality has belonged to your life. For you, it was just darkness and timeless nonexistence.

Now envision that tomorrow your doctor tells you that you are about to fall victim to the kind of brain disease described previously. However, he is able to give you a ray of hope: You may choose between two possible treatments for this disease. The first one will terminate the worsening of your condition and put you in a state of dreamless sleep

[1] In all parts of this volume written by the editors, the personal pronoun *he* is used in a gender-neutral way, meaning "he or she."

for as long as your body will live. The second treatment will put you in a state of vivid REM-sleep dreaming. Unfortunately, the latter medicine has dangerous side effects: it is poisonous for your body and will make an end of it in less than 10 years. Which one would you choose? It seems that the first treatment is essentially a nonstarter: The condition you end up with is, subjectively, no better than coma or death; as far as lived experience is concerned, you do not exist. However, the second treatment seems to be almost as good as afterlife. As research on dream mentation shows, dream experience is for the most part very much continuous with everyday experience. For instance, Snyder (1970) concluded, from his normative study of dreams using the laboratory arousal method, that dreaming consciousness is a remarkably faithful replica of waking life and that the complex visual imagery described was a realistic facsimile of the visual perception of external reality. Hence, although your memory, critical thought, or ratiocination might not work as well as in waking life, you would still in a very important sense *exist*. The brain continues to construct a world model from internal excitation during sleep, and this gives you an opportunity still to enjoy, for years to come, many of the precious conscious experiences that your life is all about.

All in all, to have conscious awareness means that, as a subject, you exist and are alive. Take away the lived experiences and you take away your life, whether or not your body continues to march on without any purpose. Give the experienced world, and you get back your life, no matter whether or not your body lies dormant and inactive, deep in sleep and isolated from external stimuli. What goes on inside your brain seems to be sufficient to support your conscious life.

Despite all this discussion to the effect that consciousness is an obvious fact of life and highly important for each and every one of us, philosophers and scientists are very confused about it. They have no agreement on what consciousness exactly is, and some are not even convinced that it exists at all, much less that it is of any importance. It is obvious that there is no unique definition or characterization of consciousness, which, of course, increases confusion and muddies the waters in scholarly discussion. For example, neuropsychologist Andrew Young (chapter 8, this volume) distinguishes among four senses of "awareness" to avoid confusion of the distinct meanings. However, it seems that the different varieties of consciousness happily coexist in our lived experience: We have the 3-D perceptual space surrounding us, the sense of the present moment, our thoughts, recollections, intentions, and action plans accompanying us wherever we go. Although some researchers prefer to attack only one aspect of experienced consciousness, the thing we really would like to know is, of course, what makes any of them instances of conscious phenomena, and what, if anything, is the property that they all share in spite of their widely differing characteristics.

PHILOSOPHY AND COGNITIVE NEUROSCIENCE

Perhaps it would be an exaggeration to say that we have no clue whatsoever to how consciousness is possible. In fact, we do have certain facts at hand that give us a hint. For example, we do know that whatever consciousness ultimately might be, it is somehow based on the workings of the central nervous system. This is something humans have not always known: For Aristotle, it was the heart, not the brain, that was the seat of our mental life. In fact, Aristotle explicitly denied that the brain could be the sensory center. For centuries, the brain was considered to be a relatively unimportant organ, the main purpose of which was to cool the blood. By contrast, at present we know that the brain is the most complex chunk of ordered material we have encountered in the universe so far. Consequently, perhaps it is not such a great miracle that that sort of a setup can give rise to something as peculiar as consciousness. Had we found, instead of the brain, just a big potato or a collection of stones inside our skulls, we would be in much worse trouble trying to explain consciousness than we now in fact are.

Another relatively safe bet is the claim that the way in which the world is modeled by an organism in its conscious experience has been biologically useful for the creature in its evolutionary history. This phenomenological model of the world has improved the likelihood of the organism's survival and the production of offspring more than other kinds of phenomenological organization have done or the absence of conscious experience could have done. Although it is difficult to trace the evolution of consciousness as long as we do not know the neurobiology on which it is based, it is encouraging that even ethologists have started to take the animal's point of view and talk about their mental states and consciousness (e.g., Cheney & Seyfarth, 1990).

Why do we, in this volume, turn to philosophy and cognitive neuroscience in our search of the secrets of consciousness? Unfortunately, we do not yet have a specific science of consciousness, so we cannot turn to "conscientologists." Psychologists, admittedly, study many conscious phenomena, but there is no explicit account of consciousness in that discipline either (cf. Farthing, 1992). So we must seek help in the field that is called *cognitive science*: the joint effort of philosophers, psychologists, neuroscientists, and artificial intelligence researchers to uncover the mechanisms of minds and intelligent systems in general.

At the moment, philosophers have an especially important role to play in the study of consciousness. They are in the habit of stepping a few steps back from the somewhat confusing trees and bushes of experimental research in order to see the whole wood of scientific knowledge. They often start off with some tacit assumptions and try to construct a coherent, all-embracing world picture from those. There are probably as many kinds of world pictures as there are philosophers, but the interesting question is: What is the place of consciousness in the order of the world, according to the philosophers and scientists of our time? What are the world pictures they offer us, and which kinds of implicit suppositions are they based on? Which of these places assumed for consciousness are the most consistent with the rapidly accumulating empirical data from other fields of cognitive science?

Thus, it is the philosophers' bold task, or perhaps burden, to sketch the answer to the most fundamental questions concerning the ultimate nature and place of the experienced realities of consciousness. Remember that we no longer have any parallel task for philosophers in the question of the ultimate nature of life. There are no opposed orthodoxies, one of which would be the claim that life is an algorithm computed by the nucleus of the cell, some other that it is a mysterious vital force, and a third one claiming that it is firmly based on biochemistry. We know that the latter view is an unchallenged doctrine of modern biology, and philosophers have, during one generation, swallowed that fact.

By contrast, in the question of the ultimate nature of mind and consciousness, the gaps between different philosophical schools of thought are still wide open. Even though everybody agrees that mind has something to do with the brain, there still is no general agreement of the exact nature of this relationship. Next, we take a short look at the history of consciousness in philosophy and in neuroscience. As this brief review shows, these two disciplines have gradually found fruitful points of contact, taking us to where we actually stand today with regard to the riddle of consciousness.

CONSCIOUSNESS AND PHILOSOPHERS: FROM ARISTOTLE TO WILLIAM JAMES

The concept of consciousness is of relatively recent origin. The philosophers of antiquity or the Middle Ages did not speak of consciousness but of soul (*psyche* in Greek and *anima* or *mens* in Latin). For them, soul did not refer to consciousness but rather to the principle of life. The first known philosopher, Thales of Miletus, seems to have thought that psyche means both life and the source of motion.

Aristotle's view was similar to Thales's but different from ours. If we ever use the word *soul* at all, it means more or less the same as consciousness, but for Aristotle soul meant that which makes a thing a living thing; that is, a principle of life (e.g., the metabolism of a plant is its psyche). Different living beings have different psychic powers: (a) Plants have the nutritive faculty only; (b) animals have the faculties of sense perception, desire, and locomotion; and (c) finally, man has reason and thought and is consequently at the top of the hierarchy of living beings.

Thus, according to Aristotle, it is rationality rather than consciousness that differentiates man from beast. He might have admitted that all living things have a degree of consciousness in their psyche, but he would have added that it is a question of complexity whether it amounts to rationality in individual cases. Aristotle's view has been aptly called the first version of functionalism. His view can be expressed functionalistically by saying that the mind of a being X is its abstract functional organization, its mode of functioning. In plain words, this means that the psychic states of a living being are the states that are used to explain its behavior. Thus, from a functionalist's point of view, a soul is a principle of explanation and not a distinct entity. This is precisely what Aristotle meant when he said that the soul is the form of the body and not an independent entity.

If we leave out early Christian theology and its ponderings on immortal souls, the 17th-century philosopher Descartes is the next thinker worthy of notice in this very brief history of consciousness. It is he who made consciousness a prominent topic in philosophy—although he himself did not use the word *consciousness* at all; instead, he spoke of "the thinking thing" (*res cogitans*) or even of soul as his predecessors had done. Indeed, the word *conscientia* does not occur even once in the Latin text of the *Second Meditation* where Descartes tried to prove that he is a thinking thing, an unembodied soul, an individual unit of consciousness. However, from the absence of the word consciousness we may not infer the absence of the concept of consciousness in Descartes' thought. The fact is that in the *Meditations* he forged the modern concept of consciousness but lacked a word to express it! (He could not use *conscientia* in the Latin text or *conscience* in the French version, because both these languages fail to make a distinction between consciousness and conscience, that is, between awareness and an inner moral voice.)

Descartes held that there are two types of substances in the world: extended things like the human body and thinking things like the human mind or soul. Thinking is the defining characteristic of man, and the ego is a thinking thing, totally different from extended physical things. Nevertheless, mind and body come together and have causal interaction at the pineal gland, the mysterious organ where this otherwise unexplainable meeting of the two substances can take place. Thus, if Aristotle is credited with the honor of being the first functionalist, Descartes is known for articulating the classical dualistic position with respect to the mind–body problem. Although there had been dualists before Descartes (e.g., Plato), Descartes was the first to make the important point that if we have two totally different substances, we are bound to have a mysterious meeting place. This holds true to this day; if the division between consciousness and the brain is made too sharp, we end up with a like situation.

Spinoza, writing a few decades after Descartes, firmly opposed Cartesian dualism. Most notably, he rejected the idea of a soul distinct from the body and yet somehow capable of causal interaction with it. It seems that he was facing a difficult dilemma: Either (a) the mind and the body are two distinct entities, as for Descartes, in which case causal interaction between them appears to be impossible, or (b) they are not distinct entities, in which case there can and must be causal interaction between what we call mind and body. Spinoza did not like either alternative. Instead, he went between the horns of the dilemma by arguing that mind and body are one and the same thing looked from two different viewpoints. This means that although every event (or state) in a human being is both mental and physical at the same time, there are two distinct orders of description and of explanation: the mental and the physical ones, each using its own terminology. Consequently, it would be a mistake to explain a mental event by means of a physical event, and vice versa, for that would amount to making a category mistake, confusing two different kinds of explanation. Yet both orders of explanation are entirely valid and respectable in themselves; they just should be kept separate, otherwise difficulties will ensue. To put Spinoza's insight in a nutshell, if a mental event is identical with a physical event, there can be no causal connection between them, for a thing just cannot be its own cause or effect. In this sense, Spinoza can be considered the forerunner of the modern *token-identity theory*, to which we return later.

Perhaps the most profound and articulated theory of consciousness ever presented is the one by William James at the end of the 19th century. His psychological theory is based on the central importance of consciousness. Our folk psychology holds that feelings can cause other mental states and even bodily actions. Any theory that denies mental causality flatly contradicts this axiom of folk psychology. James proposed that scientific psychology should be naive: It should describe the world just as it appears to us and accept the commonsense notion of mental causality. Psychology would have to pay a high price for dropping the commonsense notion of causality, and it would gain almost nothing in exchange. James also presented positive evidence for the causal role of consciousness. This is an argument based on evolution: The very existence of consciousness is intelligible only on the assumption that it is causally effective. The human animal or its brain alone could not succeed in handling the incoming information. There must be some structures that are able to pick out the useful information that in turn, help the human animal to attain well-being.

How are we supposed to study consciousness? By means of observation, James replied, but because the states of consciousness are inside the mind, they are observed introspectively, which means "looking into our own minds and reporting what we there discover." What is to be found there are states of consciousness, feelings and thoughts. James, however, departed from the tradition of those introspectionists who claimed that inner observation is infallible, a means of securing indubitable knowledge. On the contrary, introspection is like any other observation, where there is always a possibility of being mistaken or of drawing hasty, ill-founded conclusions.

What about unconscious mental events? They are not experienced and are thus out of the reach of introspection. Do they exist at all? James argued that they do not exist and supplied grounds for this apparently counter-intuitive claim. For example, habitual or automatic performances, presumably steered by unconscious mental events, could be consciously performed but so quickly that no memory of them remains. Another possibility is that the consciousness that steers the performances is cut off from the main consciousness. The possibility of multiple loci of consciousness is readily admitted by James. It is not far from admitting unconscious mental events, because the performances of one consciousness are always more or less unconscious from the viewpoint of another locus. Nevertheless, James believed that unconscious idea invites a contradiction: An idea is a representation, and it needs an onlooker, the mind's eye. Consciousness must accompany every idea—otherwise it would not be an idea. Spinoza expressed the same thought by saying that every idea is at the same time "idea of idea" (*idea ideae*) that means that every idea involves besides a representation of its object also a representation of itself; and that amounts to saying that every idea is conscious of itself.

PHILOSOPHY OF MIND WITHOUT CONSCIOUSNESS: FROM BEHAVIORISM TO MODERN COGNITIVE SCIENCE

The rise of positivism as the generally accepted philosophy of science led, in psychology, to the doctrine of behaviorism, and thus the study of consciousness entered its Dark Ages during the first decades of the 20th century. For behaviorists, psychology was first and foremost the science for predicting and controlling the behavior of organisms. The observational data of behaviorist psychologists were strictly confined to physical input stimuli and physical behavioral outputs: movements of the extremities of the body or uttered noises. Behaviorists did not bother themselves with any inner occurrences, be they mental or neurophysiological. Such phenomena, if they existed, were not proper objects for psychological investigation. Subsequently, no discussion concerning "mental states" was allowed for the behaviorist: The organism was merely a "black box," the behavior of which was to be predicted on the basis of physical stimuli. In analytical philosophy of mind, behaviorism was embraced by analyzing statements concerning mental states as statements referring to actual or dispositional behavior in response to external stimulation. All in all, behavior was not a roundabout route to study phenomena hidden behind it—such as mind, consciousness, or brain. Behavior was studied for its own sake: It was what psychology was all about. This general view of the place of psychology among other sciences and of its subject matter brainwashed the foolhardy behaviorists to believe that the time had come when psychology could discard all reference to consciousness.

Another attempt to make psychology a respectable natural science was launched in the 1950s by such philosophers as U. T. Place and J. J. C. Smart. The basic idea behind this program was that the special sciences and the entities they talk about can ultimately be explained in the theories of physics and consist of nothing but entities in the best theories of physics. This idea of the unification of science had its roots, like behaviorism, in the positivist philosophy of science. In general, this ideology may be called *reductionism* (and this is exactly what most nonphilosophers usually mean when they talk about reduction of the mind to the brain), although philosophers often call it *type-identity theory*. According to this theory, mental states should be individuated by referring to their common materialistic or neurophysiological basis. Certain types of mental states (e.g., feeling pain, seeing green, or hearing a sound) are identical with particular, as-yet-unknown states or processes of the central nervous system. In this view, pain is the firing of C-neuron fibers, and seeing green could be the firing of some other type, say, Z-neurons.

Identity theorists maintain that psychology and psychological concepts will, in the future, be reduced to neurophysiology and thus the unification of science will take one great step. As water is really H_2O, or light is really electromagnetic radiation, so hearing the sound of a harpsichord is really the firing of X-neurons, and tasting a wild strawberry is really the firing of some other neurons. Type-identity theory presupposes, then, that behind our everyday typology of mental states, there is a more fundamental, materialistic world-order, and both of them carve nature at the same joints—only the level and the language of description are different. Unlike behaviorism, type-identity theory allows us to look inside the organism and its brain and to use concepts apparently referring to mental states. Nonetheless, in the final analysis, the mental is found to be "nothing but" strictly identical to the neurophysiology of the brain. Thus, if we follow this view, consciousness ought to be identified with some as-yet unknown neural system, say, firing of XYZ-neurons. However, this prospect was not met with universal acceptance because it seemed that you cannot identify something mental, like consciousness, with something nonmental, without already leaving out the mental.

There was another line of argument against type-identity views, from which eventually emerged the next overall view of mind–brain relationships, namely *functionalism.* Functionalists pointed out that if one accepts type-identity theory, the identification of mental states with neurophysiology has to be rigorous. That is, if water is identified with H_2O, then anything that is not H_2O—no matter how much it looks, tastes, or feels like water—is not water, but something altogether different. In psychology, consequently, if consciousness is identified, for example, with firing of XYZ-neurons, then no creature without exactly those kinds of neurons can be said to be conscious, no matter how much it seems to be so. Identity theories are highly "species-chauvinist": If the identification of consciousness with human neurophysiology one day takes place, then any other creature with different neurophysiology will forever be denied to have consciousness. This kind of result seems to be exceptionally counterintuitive: We would probably like to claim that apes, monkeys, dolphins, and intelligent extraterrestrials can have the same kinds of mental states as we do, although they might have a neurophysiology different from us. Consequently, mental states are likely to be multiply realizable and not rigidly identified with any specific neurophysiology.

In functionalism, mental states are individuated by pointing to their causal, that is, functional role. Consciousness, for a functionalist, ought to be identified as that state that has certain definite causal relations to the input or sensory stimulation, other definite causal relations to other mental states (like memories), and also definite causal relations to the output, that is, the behavior of the organism. Therefore, the functionalist can say that anything that has exactly these causal relations to input, output, and other states, is consciousness: It is individuated by and identified with the realization of a set of causal relations. Note especially that functionalism is not interested in the materials that in fact realize the causal role—any causal role is multiply realizable in different sorts of materials. Every instantiation of a mental state is, in this analysis, material, but the materials themselves have nothing in common that would unite all instances under the same category. The unifying factor is the causal role of those things, not their physical nature. So, what counts also in the case of consciousness is the causal role, not the neurophysiology. Hence, every being realizing the right sorts of causal roles in its mental life has precisely the same kinds of conscious states as we do—no matter whether they are instantiated in neurons, silicon-chips, or ectoplasm. Every instance of a conscious state is *some* kind of a physical state, but they need not be the *same* kinds of physical states in all creatures or even in the same creature at separate moments. This specific version of physicalism is frequently called token-identity theory.

The initial level of functionalist explanation is the commonsense level of beliefs, desires, fears, pains, and thoughts. Commonsense psychological explanation aims to show how mental states cause other mental states and behavior. Underneath this folk-psychological level, we can use a more scientific variant of functional analysis. It can be called *subpersonal cognitive psychology* (Dennett, 1987). The general scheme here is some sort of "reiterated behaviorism" (Wilkes, 1989): The black box that represents the organism is opened, but only to place more smaller black boxes inside. Every smaller box has its own causal connections from the external world or from other boxes to other boxes or behavior. This functional decomposition can go on inside every box, always identifying more elementary functions that are collectively in charge of the complicated total behavior of the system. At the basic level the functions are so rudimentary that there is promise of uncovering the neurophysiological implementations for these functions.

Several functionalists think that, at bottom, the functioning of the whole subpersonal box model of cognition can be characterized with the help of, and also actually *is* composed of, abstract mathematical algorithms that are computed by the brain. So, there is a sort of "computer program of the mind," and if we just could find out what it is, we could program any general-purpose computer to have a mind. This kind of functionalism is frequently called *computer functionalism*, *computationalism*, or *strong artificial intelligence*. These ideas were born together with artificial intelligence in the 1950s, when it was suggested that the ontological relation between a computer and its program is the same as the relation of brain and mind. Computer programs are defined at a completely abstract level that is entirely independent of the physical features of the machine on which the program happens to run.

Unfortunately, as far as consciousness was concerned, functionalism still seemed to leave it out. Philosophical arguments against functionalism tended to concentrate around the experienced qualities of conscious mental states such as the smell of the rose or the experience of seeing colors. The point of these arguments was simply to show that a creature could conceivably have exactly the same functional characterization as you do, but nothing in this characterization reveals whether its color experiences, for example, are similar to yours. It could in principle see what you call red the way you see green, or it might not have any experiences of colors at all, although it was behaving as if it did have. Thus, the qualitative aspects of conscious experiences could be either inverted or even completely absent: Nothing in the functional organization of behavioral systems could reveal which systems are enjoying which kinds of experiences or, indeed, whether or not they are enjoying any experiences at all. It seems that two mental states could play exactly the same functional role despite the fact that only one of them was accompanied by qualitative conscious experiences. Consequently, there is a very important aspect of mentality, namely conscious experience, which a functionalistic characterization cannot capture.

Nevertheless, functionalism is the prevailing philosophy of mind and a strong undercurrent behind the new interdiciplinary research program called *cognitive science*. The great merit of cognitive science is, undoubtedly, that it has allowed a new kind of interaction to take place between psychology, philosophy, artificial intelligence, and neuroscience. The great danger cognitive science entails is that, as yet, its philosophical foundations and implications are far from clear, and it just might turn out that it treats consciousness as badly as did its predecessors.

The general idea in cognitive science has been that cognition consists of mental representations and computational processes that transform such representations to others and, in that sense, "process information." This view is very close to the way mental processes are conceptualized in artificial intelligence and in the simulation of psychological processes. Both have as their philosophical ground the functionalist theory of mind and both seem to explore psychological processes at an abstract level that does not call for analyses of brain neurophysiology or computer hardware. However, recently there has been a lot of discussion about how the computations actually are implemented in the brain, and so-called *connectionism*, which uses a "brain-style" architecture in computing, has become increasingly popular.

Nonetheless, cognitive science is still often characterized or defined as the study of human thought or intelligence in terms of *physical symbol manipulation* or *computation* (e.g., Garfield, 1990; Pylyshyn, 1989; Simon & Kaplan, 1989), and artificial intelligence is still considered to be its "intellectual core" (Boden, 1990).

THE RETURN OF CONSCIOUSNESS IN PHILOSOPHY AND NEUROSCIENCE

Although philosophers never abandoned talk about consciousness quite as completely as did experimental psychologists, consciousness was rarely treated explicitly as the main topic during the murky days of behaviorism. Identity theorists, by contrast, unashamedly asked the question, "Is Consciousness a Brain Process?" (Place, 1956). Functionalist philosophers of mind, for their part, tended to hide the problem under the stairs, once again, but it kept appearing in varying guises, reminding thinkers every now and then that the mind–brain relationship was far from resolved. The three most prominent manifestations of the problem were the recurrent discussions and arguments about *subjectivity*, *qualia*, and *intentionality*.

Perhaps the most remarkable landmark of the return of consciousness and subjectivity as the main focus of the philosophy of mind was the publication of Thomas Nagel's classic article, "What Is It Like to Be a Bat?" in 1974. In this article, Nagel analyzed what it means for an organism to have conscious experiences: "An organism has conscious mental states if and only if there is something it is like to *be* that organism; something it is like *for* the organism" (Nagel, 1974, p. 436). Moreover, he claimed that the various attempts to reduce mental states to this or that had all been unsuccessful, because they all ignored this subjective character of conscious mental experiences.

By qualia, philosophers mean those properties of conscious experiences that define what the experience feels like for the subject. The taste of chocolate, the itch of a mosquito bite, the heat of the sauna, the shrill, chirping noise of a grasshopper, and the pale, yellowish glow of the full moon are qualitative properties of conscious experiences. The external physical causes of such experiences, the behavioral responses associated with them, or the neural impulses in the brain underlying them seem to be only accidentally connected with the experiences themselves. The qualia, by contrast, seem absolutely essential for the conscious mental states to exist at all. As we already mentioned previously, numerous arguments against functionalism were based on the impression that the theory completely leaves out the experienced qualities of conscious mental states and, consequently, can never amount to a sufficient and satisfactory account of the mind. (For classical qualia arguments, see Block, 1980a, 1980b; Jackson, 1982.)

The third important connection of functionalistic philosophy to consciousness came through intentionality, which was perhaps the central problem of the philosophy of cognition throughout the 1970s and 1980s. This was because the central concepts of cognitive science—like representation, belief, and perception—were intentional notions in the sense that such states are directed at objects and states of affairs in the world. The main task was to determine how intentional states can stand for something outside of themselves; how they can have content or semantics. The philosopher John Searle (1980) made a distinction between *intrinsic intentionality* and *observer-relative ascriptions of intentionality*: Only mental states can have intrinsic intentionality, but beings with intrinsically intentional mental states can ascribe observer-relative intentionality to any systems they like to treat as if having beliefs, perceptions, or representations. However, true intentionality has, according to Searle (1979), a necessary connection to consciousness: Only beings capable of conscious states can have intrinsically intentional states. Thus, Searle's idea was that only conscious beings can really perceive, understand, represent, or believe anything. Nevertheless, most of the work on intentionality had been made completely disregarding its connections to consciousness, and later Searle (1989, 1990, 1992) launched a major attack against cognitive science ignoring consciousness.

So, functionalism was not quite capable of drowning consciousness for good—it kept stubbornly resurfacing in the clothes of subjectivity, qualia, and intentionality. Meanwhile, in the department of neuroscience, a number of significant experiments resulting in surprising findings were made. These investigations touched on human consciousness so closely that in consequence of them many neurologists and neuropsychologists were virtually compelled to become amateur philosophers of mind. Gradually, rumors of such new alchemy reached the department of philosophy as well. The most famous experiments on consciousness in neuroscience include the direct excitation of the cortex, split-brain studies, the phenomenon of blindsight, and measurements of the timing of conscious experience.

The brain surgeon Wilder Penfield made a series of experiments in the 1950s by directly stimulating the exposed brains of conscious epilepsy patients undergoing brain surgery. Electrical stimulation of certain cortical areas caused many unexpected effects: Sometimes they presented to the patient various crude sensations when applied to the primary cortical reception areas; sometimes they caused the patient to turn his head and eyes, or to move his limbs, or to vocalize and swallow. They also generated illusions to the effect that the present experience is familiar, or that the objects they see grow larger and come nearer.

The most startling effect was that electrical stimulation of the temporal lobe caused several of the patients subjectively to relive some past experiences: They reported vivid visual and auditory imagery, while they remained aware of the actual surroundings. These curious playbacks of past episodes were not described as "rememberings"; they were better characterized as dreams or flashbacks of the person's actual past life. For instance, a mother experienced being in her kitchen and listening to the voice of her little boy and the noises of outside traffic. A man had the experience of attending a baseball match, watching a boy crawl from under the fence to join the audience. Another patient experienced being in a concert hall, listening to music and hearing all the different instruments. One patient experienced meeting his far-away friends and laughing with them. All these stimulation-elicited replays of past experiences seemed realistic, while at the same time the patient remained aware of the operation room. One patient explained that she saw "the people in this world and in that world at the same time." (For a review of these "experiential responses to stimulation," see Penfield & Perot, 1963.) These findings created many questions concerning the relationship between the neuronal records stimulated and the subjective experiences thus elicited. Penfield (1975) himself eventually turned to dualism and speculated about "mind energy" and direct communication with the mind of God.

In the 1960s, another kind of evidence started to accumulate in neuropsychology that raised considerable speculation concerning the nature of consciousness. In short, it seemed very likely that a surgical division of the brain by the midline section of the cerebral commissures entailed a co-occuring division of the mind and consciousness. The cerebral hemispheres of these split-brain patients were separated from each other by sectioning the corpus callosum and other commissures that normally combine the hemispheres. Usually this was done in order to tame an unbearably violent case of epilepsy. After the operation, the patients normally recovered well, and nothing unusual could be noticed in their everyday behavior.

However, in controlled laboratory tests split-brain patients manifested surprising dissociations of mental processes. A common experimental setting was something like this: The patient was asked to fixate his gaze onto the center of a screen. After that, two words were briefly flashed on the screen so that one of them was to the left of the patient's fixating point and the other to the right. The exposure time was so short that it was impossible for the subject to scan the words with several fixations. Thus, it was ensured that from the left visual field information goes only to the right hemisphere and from the right visual field to the left—the projection areas are always in the contralateral sides of the brain.

Let us suppose that the flashed words were *toad* (to the left visual field and thus to the right hemisphere) and *stool* (to the right visual field and thus to the left hemisphere). Now, if the subject is asked what he saw, he answers that he saw the word *stool*. The speech-production mechanisms are, for most people, exclusively in the left hemisphere—hence, that hemisphere has only seen *stool*. No matter how you question him, you get no verbal answers (although you might get an emotional reaction) that would reveal that he had had any awareness whatsoever about *toad* or *toadstool*. Nonetheless, if the patient is asked to point with his *left* hand—which is almost exclusively controlled by the right hemisphere—to a picture representing the word that he saw, he will point to a picture of a frog-like animal, if the other choices are a seat without a back and an umbrella-shaped fungus. It seems, then, that we have one subject fully aware of the concept of stool (a seat without a back), one fully aware of the concept of toad (a frog-like animal), and no subject who has seen the word *toadstool* and who would, consequently, be aware of the concept of an umbrella-shaped fungus.

Some patients do have crude language-producing capacities also in their right hemispheres. When the separated hemispheres of one such patient were asked the same questions, they did not give identical answers. When asked which job he would choose, one patient responded with his left hemisphere "draughtsman," but, by contrast, the right chose "automobile race." Findings like this have led to controversy concerning the unity of consciousness, and it has been suggested that the patients possess two independent conscious minds, although competing interpretations also exist, counting the patients' minds differently.

Split-brain research is a fine specimen of astounding empirical studies that made neuroscientists philosophize about consciousness and after a time-lag of some years for the news to spread, forced philosophers of mind to get acquainted with actual experimental settings and results. No general agreement was ever reached: Almost every possible interpretation of the phenomena had its own supporters. Thus, the pioneering neuropsychologist and Nobel laureate Roger Sperry argued that there are two independent minds, two actual streams of consciousness flowing inside the split brain (Sperry, 1984). Another researcher in the field, Michael Gazzaniga, was more inclined to allow conscious status only to the verbally behaving left hemisphere (Gazzaniga, LeDoux, & Wilson, 1977). But philosophers joined this game of mind counting and came up with even more incredible answers. Thomas Nagel (1979) concluded that "there is no whole number of individual minds that these patients can be said to have" (p. 163). Charles Marks (1981) favored a one-mind account of split-brain patients, although, he claimed, a two-minds account could not be rejected on logical grounds alone. Roland Puccetti (1981) argued that, in fact, each hemisphere has a mind of its own not only after commissurotomy but also normally for all of us, all of the time. Patricia Smith Churchland (1981) criticized this whole debate, pointing out that the paucity of theories concerning consciousness makes it difficult to know what the discussion is all about. She suggested that "being conscious" is at this stage of mind–brain science theoretically so ill-defined that counting centers of consciousness in the brain is a bit like asking how many angels can dance on the head of the pin or how many vital spirits pervade living organisms.

The third empirical finding that has slowly caused the raising of eyebrows among philosophers was originally reported by Ernst Pöppel and his colleagues in 1973, followed in 1974 by Larry Weiskrantz and his colleagues. The initial reports revealed that patients who apparently had blind scotomas in their visual fields resulting from brain injury of the primary visual cortices, showed the unexpected capacity of *residual vision* or, as Weiskrantz aptly christened it, *blindsight*. This meant that the patients, despite their evident blindness, were able to direct their eyes to where a stimulus was located in the "blind" field. Later it was found that such patients were also able to point to unseen stimuli and even to make discriminations between various kinds of stimuli. In these experiments, the patients had to be persuaded to guess or to make forced choices between alternatives, because it was difficult to convince them that the task was not irrational. The patients felt that they were just making random guesses out of the blue and they were thoroughly amazed when told about the success of their responses. Phenomenologically, the patients do not *see* anything—they have no visual experiences, but somehow the behavioral responses are influenced by the unseen stimuli.

Blindsight proved to be just the forerunner of a whole bunch of phenomena, nowadays grouped together under the labels *implicit knowledge* or *knowledge without awareness* in neuropsychological syndromes. Thus, a considerable variety of memory deficits and recognition or perceptual impairments have been shown to manifest a pattern roughly similar to blindsight: Although the patient subjectively lacks a certain cognitive capacity and, more importantly, a certain type of conscious experience, his brain can nevertheless be shown to exhibit many kinds of discriminative responses that the patient is completely unaware of and cannot utilize in his everyday behavior. Consequently, neuropsychologists have tried to extrapolate the role of normal consciousness in cognition from the fact that consciousness can be disconnected from cognitive processes in these specific but unexpected ways (Schacter, McAndrews, & Moscovitch, 1988).

Benjamin Libet and his collaborators repeatedly presented findings about the timing of conscious experiences and their neurophysiological correlations in the brain. In 1979, they compared the way in which we experience somatosensory stimuli applied either to the skin or directly to the somatosensory cortex. They first found out that a cerebral stimulus has to be applied for about 500 milliseconds before it produces any conscious experience. If the duration was shortened 10%–20%, the subjects were confident that they had not experienced it. They hypothesized that a stimulus applied to the skin and traveling to the somatosensory cortex would, accordingly, require a similar period of about 500 milliseconds for "cerebral neuronal adequacy" to develop, eliciting the conscious experience. However, when the cortex was stimulated first and the skin next—less than the critical 500 milliseconds later—the subjective order of the stimuli was, strangely enough, reversed: The subjects reported experiencing the skin stimulus first (although it was applied last) and the cerebral stimulus last (although it was applied first). Libet's controversial interpretation was that because stimuli applied to the skin and stimuli applied straight to the cortex need, as far as we know empirically, about 500 milliseconds to be consciously experienced or reported, the skin stimulus was somehow referred backwards in time so that it seemed to the subject as if the skin stimulus was experienced first and the cortical after it. This finding, according to him, seems to raise "serious though not insurmountable difficulties" (Libet, Wright, Feinstein, & Pearl, 1979, p. 222) for the hypothesis that the psychological is identical with the neural. Later Libet (1985) was interested in the neurophysiological correlates of voluntary, spontaneous actions and, thus, the physiology of conscious free will. He found that even the most spontaneous deliberate movements and decisions to move were preceded by so-called "readiness potentials" about 200–400 milliseconds before the actual experienced intention to move. This finding, Libet insisted, had implications for what we should think of our free will and conscious intentions to carry out actions. Libet's and his colleagues works caused and still continue to cause both uneasiness and applaud among neuroscientists and philosophers, the most recent examples of the former being Dennett (1991) and Dennett and Kinsbourne (1992).

In a nutshell, consciousness survived in philosophy in the form of the annoying problems of subjectivity, qualia, and (intrinsic) intentionality. At the same time, it started to raise its head also among neuroscientists as a consequence of curious experimental results that demanded theoretical interpretations in terms of subjective experience.

PHILISOPHY AND COGNITIVE NEUROSCIENCE NOW: A JOINT EFFORT TO UNRAVEL THE WORKINGS OF COGNITION AND CONSCIOUSNESS

In 1949, the philosopher Gilbert Ryle was still constructing a wholesale "theory of mind" that was entirely an armchair product, concerning the application of concepts with which we describe the mind—the mind's "logical geography." Twenty years later, however, Daniel Dennett was already convinced that philosophers could not go on studying the mind in a factual vacuum. In the preface to his 1969 book, *Content and Consciousness*, Dennett started to close the gap between the empirical and the conceptual inquiry of mind, observing that attempts to answer many pressing questions about the mind–brain relationship had been confined "to philosophical guesswork on the one hand and the speculative perorations of retiring professors of neurology on the other" (Dennett, 1969, p. ix). Thus, Dennett's approach was among the first to cross the disciplinary gaps; a feature that has become distinctive of explorations in cognitive science.

By the 1960s, neuroscientists had also noticed the same gap: the famous neurologist Lord Brain declared in 1963 that the time was past when philosophers could ex cathedra issue naive views on "brain and mind" or reality and appearance without penetrating the physiological aspects of these problems in detail. A similar warning against physiologists tinkering naively with philosophical questions had already been made in the 1950s by the neuroscientist Ragnar Granit, whom Brain was quoting.

Dennett continued on the path he had chosen and in 1978 published *Brainstorms*, a collection of papers that constituted a theory of mind. In the introduction to that volume, Dennett's tone of voice is more optimistic than in his previous book, "Times have changed. Psychology has become 'cognitive' or 'mentalistic' . . . and fascinating discoveries have been made about such familiar philosophical concerns as mental imagery, remembering and language comprehension. Even the brain scientists are beginning to tinker with models that founder on *conceptual* puzzles" (Dennett, 1978a, p. xiii).

Others followed. Jerry Fodor's (1983) monograph, *The Modularity of Mind*, was an excellent exercise in "speculative psychology" that had profound implications to both empirical cognitive psychology and philosophy of mind. A final cornerstone to the new approach was

Patricia Smith Churchland's (1986) book, *Neurophilosophy*, in which, probably for the first time, cognitive, neuroscientific, and philosophical inquiry were merged together. Churchland's declaration was that "it is now evident that where one discipline ends and the other begins no longer matters, for it is in the nature of the case that the boundaries are ill defined" (Churchland, 1986, pp. ix–x).

At that time, it was also becoming more and more evident that consciousness was the central but as yet largely untouched problem for such interplay of cognitive science, neuroscience, and philosophy. Gradually this has been realized, and since Dennett's (1978b) paper, "Toward a Cognitive Theory of Consciousness," we have seen theories or models of consciousness appearing with increasing frequency. Perhaps one publication needs to be mentioned separately: the collection of papers edited by A. J. Marcel and E. Bisiach (1988), *Consciousness in Contemporary Science*. In the introduction to that book, the editors noted that "psychology of various kinds, neurology, neurophysiology, artificial intelligence, and philosophy are interacting at the moment in a way they did not before" (Marcel & Bisiach, 1988, p. 6). And they suggested that the crux of their book is whether consciousness poses a problem for functionalism, the common core underlying cognitive inquiry. There we had, for the first time, the empirical and the philosophical community together focusing on consciousness. That was a major landmark in the study of consciousness, showing us the course we still consider the most reasonable to take.

So, where exactly do we stand today? Judging from the articles and books published, we should say that consciousness has taken its place as the central problem of the philosophy of cognition. Also in experimental psychology and cognitive neuropsychology there is a tendency to discuss the results in terms of awareness and consciousness; the sterile notions of "controlled processing," "working memory," or "explicit knowledge" perhaps have the virtue of not mentioning anything as weird as consciousness, but because that is largely what they are talking about, why should we pretend? The serious study of consciousness even has its own journal, *Consciousness and Cognition*, that was started in 1992. What, then, might we still be missing?

The difficulty, it seems, is that there is no agreement in the scientific community on what it is that we are actually studying when we do research on consciousness. This, of course, is a radically different situation from that in biology, for example. The subject matter of biological research is life, and there are no profound disagreements among biologists all over the world what sort of a phenomenon life in fact is. The average biologist does not really have to worry about the philosophical debates concerning the nature of life and living matter, because they have been, for all practical purposes, settled decades ago. In the mind sciences, by contrast, the situation is not that happy. After all, the subject matter of psychology is, to a great extent, consciousness. If an average experimental psychologist uncovers some results that invite him to talk about consciousness, he is in deep trouble. There is no one who could tell him what consciousness is, no textbook definitions; only an unending philosophical quarrel. Ask one philosopher and he says that consciousness is a virtual Von Neumann machine realized in a connectionist architecture, computing computable functions. Ask the next, and he replies that the concept of consciousness should not be included into psychological theories: it is too vague and probably refers to nothing at all. The third might believe that consciousness is a biological phenomenon like any other, produced enigmatically inside the neurobiological machinery of the central nervous system. Some might warn that consciousness is so mysterious that you face a hopeless task if you try to unveil its secrets: Human cognition is inadequate for understanding itself. You might even encounter an outmoded dualist, trying to convince you of the mechanisms between the noble, conscious soul and the messy, earthly biology.

As it stands, nobody is going to tell you the orthodox view, because there is no such thing: you must make up your mind yourself. This situation, of course, cannot prevail, if we want to have a science of consciousness. At the moment, the science of consciousness is, at best, "the science of . . . WHAT?"

THE ORGANIZATION OF THE BOOK

The book is composed of four thematic parts. Each part begins with an introductory section, the purpose of which is to illuminate the background of the often complicated philosophical and theoretical discussions. Part 1 of the book is titled in the form of the question, "What is Consciousness?" The main focus is on the fundamental nature or as philosophers often call it, the *ontological status* of consciousness. The answer to this question, if it can be agreed on, profoundly influences the way we should think about consciousness and the way we ought to explore it in science.

Part 2, "Properties of Consciousness," introduces the problematic aspects of conscious experiences, or how science meets the fragile smell of rose, the peculiar "aboutness" of thought, and the elusive subject of conscious experience. These three problems are traditionally known to philosophers under the names of qualia, intentionality, and the subjective point of view. It seems that the way a philosopher treats these features of conscious experience is largely interconnected to how the same philosopher treats consciousness in general. The chapters in this section concentrate especially on qualia and intentionality.

In Part 3, "Models of Consciousness," the authors approach empirically based models of conscious events in the mind–brain. In the introductory section, "Chasing the Ghost in the Machine," modern models of consciousness are reviewed and evaluated in the light of their differing philosophical implications. The chapters in this Part take us deeper into two specific areas of inquiry: the global workspace theory of consciousness and the lessons we can learn from neuropsychological disorders in relation to consciousness.

In Part 4, the final part of the volume, we attempt to catch a glimpse of the "Future of Consciousness." "Will we get an explanation of consciousness?" is a fascinating question for both philosophers of mind and philosophers of science as well as for those empirical researchers who seek the unification of theories at different levels. Despite the manifest disagreement and multiplicity of differing views, the conclusion seems to be that there is light at the end of the tunnel: Not all theorizing about consciousness is futile, but the task now is to clear the waters and to seek for some common ground on which to build the future of the study of consciousness.

REFERENCES

Block, N. (1980a). Troubles with functionalism. In N. Block (Ed.), *Readings in philosophy of psychology* (pp. 268–306). Cambridge, MA: Harvard University Press.

Block, N. (1980b). What is functionalism? In N. Block (Ed.), *Readings in philosophy of psychology* (pp. 171–184). Cambridge, MA: Harvard University Press.

Boden, M. A. (Ed.). (1990). *The philosophy of artificial intelligence.* Oxford: Oxford University Press.

Brain, L. (1963). Some reflections on brain and mind. *Brain, 86*, 381–402.

Cheney, D. L., & Seyfarth, R. M. (1990). *How monkeys see the world: Inside the mind of another species.* Chicago: University of Chicago Press.

Churchland, P. S. (1981). How many angels . . . ? *Behavioral and Brain Sciences, 4*, 103–104.

Churchland, P. S. (1986). *Neurophilosophy: Toward a unified science of the mind-brain*. Cambridge, MA: MIT Press.

Dennett, D. C. (1969). *Content and consciousness.* London: Routledge & Kegan Paul.

Dennett, D. C. (1978a). *Brainstorms.* Brighton: Harvester.

Dennett, D. C. (1978b). Toward a cognitive theory of consciousness. In D. C. Dennett (Ed.), *Brainstorms* (pp. 149–173). Brighton: Harvester.

Dennett, D. C. (1987). *The intentional stance*. Cambridge, MA: MIT Press.

Dennett, D. C. (1991). *Consciousness explained.* Boston: Little, Brown.

Dennett, D. C., & Kinsbourne, M. (1992). Time and the observer: The where and when of consciousness in the brain (with commentary). *Behavioral and Brain Sciences, 15*, 183–247.

Farthing, W. G. (1992). *The psychology of consciousness*. Englewood Cliffs, NJ: Prentice Hall.

Fodor, J. A. (1983). *The modularity of mind*. Cambridge, MA: MIT Press.

Garfield, J. L. (Ed.) (1990). *Foundations of cognitive science*. New York: Paragon House.

Gazzaniga, M. S., LeDoux, J. E., & Wilson, D. H. (1977). Language, praxis, and the right hemisphere: Clues to some mechanisms of consciousness. *Neurology, 27*, 1144–1147.

Jackson, F. (1982). Epiphenomenal qualia. *Philosophical Quarterly, 32*, 127–136.

Libet, B. (1985). Unconscious cerebral initiative and the role of conscious will in voluntary action (with commentary). *Behavioral and Brain Sciences, 8*, 529–566.

Libet, B., Wright, E. W., Feinstein, B., & Pearl, D. K. (1979). Subjective referral of the timing for a conscious sensory experience. *Brain, 102*, 193–224.

Marcel, A. J., & Bisiach, E. (Eds.). (1988). *Consciousness in contemporary science*. New York: Oxford University Press.

Marks, C. E. (1981). *Commissurotomy, consciousness & unity of mind*. Cambridge, MA: MIT Press.

Nagel, T. (1974). What is it like to be a bat? *The Philosophical Review, 83*, 435–450.

Nagel, T. (1979). Brain bisection and the unity of consciousness. In T. Nagel (Ed.), *Mortal questions* (pp. 147–164). London: Cambridge University Press.

Penfield, W. (1975). *The mystery of the mind*. Princeton, NJ: Princeton University Press.

Penfield, W., & Perot, P. (1963). The brain's record of auditory and visual experience: A final summary and discussion. *Brain, 86*, 595–696.

Place, U. T. (1956). Is consciousness a brain process? *British Journal of Psychology, 47*, 44–50.

Puccetti, R. (1981). The case for mental duality: Evidence from split-brain data and other considerations (with commentary). *Behavioral and Brain Sciences, 4*, 93–123.

Pöppel, E., Held, R., & Frost, D. (1973). Residual visual function after brain wounds involving the central visual pathways in man. *Nature, 243*, 295–296.

Pylyshyn, Z. W. (1989). Computing in cognitive science. In M. I. Posner (Ed.), *Foundations of cognitive science* (pp. 49–92). Cambribge, MA: MIT press.

Ryle, G. (1949). *The concept of mind*. London: Hutchinson.

Schacter, D. L., McAndrews, M. P., & Moscovitch, M. (1988). Access to consciousness: Dissociations between implicit and explicit knowledge in neuropsychological syndromes. In L. Weiskrantz (Ed.), *Thought without language* (pp. 242–278). Oxford: Oxford University Press.

Searle, J. R. (1979). What is an intentional state? *Mind, 88*, 72–94.

Searle, J. R. (1980). Minds, brains, and programs. *Behavioral and Brain Sciences, 3*, 417–457.

Searle, J. R. (1989). Consciousness, unconsciousness, and intentionality. *Philosophical Topics, 18*, 193–209.

Searle, J. R. (1990). Consciousness, explanatory inversion and cognitive science. *Behavioral and Brain Sciences, 13*, 585–642.

Searle, J. R. (1992). *The rediscovery of the mind*. Cambridge, MA: MIT Press.

Simon, H. A., & Kaplan, C. A. (1989). Foundations of cognitive science. In M. I. Posner (Ed.), *Foundations of cognitive science* (pp. 1–47). Cambribge, MA: MIT Press.

Snyder, F. (1970). The phenomenology of dreaming. In L. Madow & L. H. Snow (Eds.), *The psychodynamic implications of the physiological studies on dreams* (pp. 124–151). Springfield, IL: Thomas.

Sperry, R. (1984). Consciousness, personal identity and the divided brain. *Neuropsychologia, 22*, 661–673.

Weiskrantz, L., Warrington, E. K., Sanders, M. D., & Marshall, J. (1974). Visual capacity in the hemianopic field following a restricted occipital ablation. *Brain, 97*, 709–728.

Wilkes, K. V. (1989). Mind and body. *Key Themes in Philosophy, Royal Institute of Philosophy Lecture Series, 24*, 69–83.

1 WHAT IS CONSCIOUSNESS?

Antti Revonsuo
University of Turku, Finland

THE APPEARANCE AND REALITY OF CONSCIOUSNESS

The purpose of metaphysical investigations is to resolve which kinds of things are real and exist in the world. Now, consciousness is a tricky subject for such inquiry, because by definition, what we have in conscious experience is often labeled as "mere appearance." This appearance is contrasted with what "really" exists in the world, beyond appearances, thus implying that conscious experience enjoys a mode of existence not quite deserving to be named "real." But how should we understand the claim that conscious experience is not real? What could it possibly mean? If we take the claim at face value, it seems to flatly deny that conscious experiences exist at all. And this indeed is how some philosophers boldly respond to the metaphysical ponderings concerning the nature of consciousness. Nevertheless, a rejection of consciousness strikes at the roots of sanity, pulling the rug from under the world we seem to live in. If our life consists of a succession of lived experiences, it is hard to make sense of a claim to the effect that those experiences do not really exist. Does it mean that all our lives we have been under some sort of mass delusion, mistakenly believing that we enjoy experiences? Those who offer the elimination of consciousness as a solution often point to the history of science where progress was sometimes made by recognizing the false ontological commitments of old theories in the face of new, better theoretical frameworks. Unfortunately, no eliminativist of consciousness has offered a superior theoretical framework that explains our lived experience without evoking the notion of consciousness and which would make the falsities in the old view understandable.

If the denial of consciousness does not make sense, what sort of reality is the appearance of conscious experience? One currently unpopular way to anchor the reality of consciousness is to postulate an other-worldly plane of existence for the substance of which consciousness and the mental are "made of." The utter inappropriateness of such a dualist solution in the modern scientific world view is all too well-known, and it only manages to get us out of the frying pan and into the fire. The mystery of soul-stuff and its interaction with the earthly brain are no less enigmatic than consciousness itself is. Well, why not start with something that is not enigmatic, something undeniably real? The brain, it seems, is as real a material object as are the sticks and stones around us. We know that consciousness has something to do with the brain, anyway, so perhaps consciousness just is a brain process. The "is" in that claim ought to be taken literally: Perhaps consciousness is identical with—nothing but—a brain process.

This view certainly has been advocated by many, but it is not without its dilemmas. How could any two different things be identical with each other? Clearly, they could not and, consequently, it is difficult to see how the mental could be literally identical with anything nonmental, without already omitting the mental from the analysis. Another inconvenience arises from the fact that, obviously, many different kinds of creatures on the earth are conscious, and possibly somewhere in the Milky Way or remote galaxies, there are creatures genuinely conscious but built on an entirely alien neurobiology. The one and same property, consciousness, certainly cannot be identical with different sorts of physical properties any more than, for example, the property of being a carbon atom could here be a matter of having six protons and six electrons, but in the Pleiades, a matter of having two protons and two electrons.

Every instance of conscious processes, nevertheless, is realized in some sort of physical matter, although it seems that consciousness itself might not be strictly identical to any specific type of physical matter. Then perhaps we must turn our attention to properties of the brain different from the purely neurophysiological ones. At the levels of the functions and causal relations of brain events we might be able to identify some special function, the realization of which is the landmark of consciousness. Whichever physical system then realizes these causal relations also realizes consciousness: We may identify it with something at the functional rather than the physical level. Moreover, it might be possible to describe such a function as computable algorithms and thus, to implement it as a computer program. In running this program, the computer then realizes the abstract causal relationships and, subsequently, enjoys consciousness. But what sort of an ontological assumption is it to say that consciousness is a computer program? It is dubious whether functionalist characterizations can capture any of the bare essentials of conscious experience such as subjective, qualitative feels or even truly intentional mental states. Furthermore, it is questionable whether it makes any sense to assume that brain processes are "computing" anything or that they even have any "functions" independent of us, the interpreters of the system.

None of the previously mentioned standard answers seem to really work when it comes to explaining consciousness. In Part 1 of this volume, "What Is Consciousness," four eminent philosophers introduce their fresh approaches to the puzzle. In chapter 1, Ilkka Niiniluoto outlines the problem of consciousness from the perspective of scientific realism. The basic assumptions behind that approach are that there indeed does exist a reality independent of consciousness and, furthermore, that it is possible for human science to gain knowledge concerning this reality. Thus, scientific realism implies just the distinction between the manifest and scientific images of the world; between appearances and reality. In the long run, the scientific image, presumably, is the more profound and ontologically more basic account of the world. But, we may ask, how does scientific realism face consciousness itself—the root of all appearances? What is the scientific image of the manifest image? What is the real nature of the appearance? Niiniluoto proposes that from the point of view of scientific realism, the interaction between philosophy of mind and scientific psychology can lead to a better understanding of the nature of consciousness. Nonetheless, remarks Niiniluoto, the metaphysical controversies between rival theoretical approaches to consciousness have not been solved by empirical research. Thus, our choice of a favorite theory of consciousness still tends to be based on our philosophical "tastes" and pretheoretical convictions.

In chapter 2, Daniel C. Dennett sets out to defend his new *multiple drafts theory* (or metaphor) of consciousness against claims according to which his theory does not admit the reality of consciousness (see, e.g., Niiniluoto, chapter 1, this volume; Churchland & Ramachandran, chapter 3, this volume; Revonsuo, chapter 11, this volume). This model was introduced by Dennett (1991) and Dennett and Kinsbourne (1992), but it has been frequently criticized for not being clear about the reality and ontological status of consciousness. Dennett claims that we still tend to think of consciousness as a centered locus of subjectivity or a point of view somewhere inside the brain. This is, in his view, a perilous remnant of Descartes' interactionist dualism, according to which there has to be a centralized gateway from the material brain to the immaterial conscious mind. Of course, practically nobody today believes that the mind is a separate, immaterial entity to which information from senses is broadcast through the pineal gland. However, even modern materialists seem to think of consciousness as some kind of material and functional center in the brain. The idea of a center of consciousness where "it all comes together" as multimodal, determinate experiences implies that the necessary and sufficient condition for a mental operation to be conscious is that the operation in question is momentarily "inside" this center. These ideas are buried very deep in our everyday intuitions concerning what kind of a phenomenon consciousness must be. Dennett argues that those intuitive ideas are, nevertheless, fundamentally incorrect, and they ought to be replaced with a model that is both philosophically and empirically more credible.

Dennet naturally has his own peculiar philosophical commitments on which the alternative he proposed is based. His metaphysical bedfellows in the philosophy of mind are, to some extent at least, Gilbert Ryle's *logical behaviorism* and V. W. O. Quine's *ontological relativism*. Ryle (1949) wanted to give the fatal blow to "Decartes' myth" or the "dogma of the ghost in the machine," the idea that minds are extra centers of nonmechanical causal processes. He subscribed to an instrumentalist methodology of science, according to which theories are mere instruments assisting us, for example, to predict and explain the observable behavior of people. Quine, for his part, argued that theories of meaning should be based on data consisting of overt behavior, which always leads to a number of different interpretations of the same data. In the final analysis, there is no fact of the matter of what people (and even theories) refer to: "Reference is nonsense except relative to a coordinate system" (Quine, 1969, p. 523). In the spirit of these philosophers, Dennett (1987) introduced his *intentional system theory*. It is a sort of *holistic logical behaviorism* and deals with the "prediction and explanation from belief-desire profiles" (p. 58) of the actions of whole intentional systems. The interpretations of a system's behavior in terms of beliefs and desires is ontologically relative in the way that there is no fact of the matter of what a system really believes: There is no right interpretation of a system's intentional or other mental states. Dennett's multiple drafts theory is an attempt to extend ontological relativism all the way to consciousness. According to Dennett (1991), "the multiple drafts model avoids the tempting mistake of supposing that there must be a single narrative . . . that is canonical—that is the *actual* stream of consciousness of the subject" (p.

113). In other words, there is no single stream of consciousness, and our contents of consciousness are never determinate. Dennett backs these claims by showing how the novel theory can handle certain puzzling empirical phenomena in which the relation between the subjective temporal ordering of conscious experiences is hard to reconcile with the temporal ordering of objectively measurable brain events that supposedly underlie those experiences. The flavor of paradox is due to our tendency to think in terms of a determinate center of consciousness. If we realize that fact, we see how the multiple drafts model is able to avoid the empirical paradoxes and the metaphysical dangers surrounding the traditional way of thinking, maintains Dennett.

Dennett's (1987) starting point is the "objective, materialistic, third-person world of the physical sciences" (p. 5), and he (1991) tries "to explain every puzzling feature of human consciousness within the framework of contemporary physical science" (p. 40). From that point of view, the contents of consciousness are indeed hard to anchor absolutely and, consequently, the relativism might hold. But how do we reconcile that view with the very different first-person view each and every one of us has of consciousness and its contents?

P. S. Churchland and V. S. Ramachandran, in chapter 3, heavily criticize some of the claims Dennett made in connection with his multiple drafts model. Churchland's objection concerns Dennett's assertations about the phenomenon called *filling in*. For exmple, there is a letter missing in one word of this sentence. Did you notice the typographical error, or did your brain "fill in" the missing 'a' in 'exmple'? Dennett denies that any kind of filling in ever really happens in our sensory experiences. He claimed that the brain merely *judges* that there ought to be something in an unperceived location of our perceptual space, for example, in the blind spot, but that the brain never creates any additional sensory experiences where the sensory input mechanisms provide no stimulation. Churchland and Ramachandran vigorously argue that Dennett's claims are not in concert with empirical findings. They suspect that the reason for Dennett's reluctance to admit the reality of filling in is the behavioristic ideology lurking behind the scenes in Dennett's philosophy.

In chapter 4, John R. Searle introduces *biological naturalism* and some of its implications to the study of consciousness. Searle's approach is undoubtedly realistic about consciousness: He thinks that we must take the subjectivity and first-person character of consciousness as "brute facts of nature" and start seeking neurobiological explanations for that peculiar but real feature of the universe. Searle argues that cognitive science has been seriously astray when it has tried to construct an explanation of the mental in terms of programs and computations.

Nothing in physical nature, independent of us, is a computer program or a computation, and thus even the question, "Is consciousness a computer program?", lacks a clear sense, claims Searle. If he is right, incorporating the study of consciousness into cognitive science may imply that many of the basic notions in that science have to be re-evaluated.

Dennett, Churchland, and Searle represent three rather dissimilar approaches to the philosophy of mind, and this fact, accordingly, shows itself in their attitudes towards consciousness. Dennett's constitutional starting point is a theory of content, that is, *intentionality.* He writes (1991), "My fundamental strategy has always been the same: first, to develop an account of content that is *independent of* and *more fundamental than* consciousness . . . and secondly, to build an account of consciousness on that foundation" (p. 457). Thus, Dennett thinks that he can give an account of "aboutness" that does not presuppose consciousness. Moreover, he subscribes to some form of functionalism and believes that "an appropriately programmed computer—provided only that it is fast enough to interface with the sensory transducers and motor effecters of a 'body' (robot or organic)—literally has a mind, whatever its material instantiation, organic or inorganic" (Dennett, 1990, p. 50).

Searle holds a completely opposing view. His starting point is the philosophy of language that he regards as a branch of the philosophy of mind. The content of linguistic elements is, in Searle's view, grounded on the basic, biological form of intentionality—*intrinsic intentionality*—that can exist only in the biological brain–mind (and maybe also in systems sufficiently alike the biological brain). Searle argues that only beings capable of having conscious states can even in principle have intrinsically intentional states: Consciousness is the basic landmark of the mental, and the existence of intentionality is dependent on it. Computer programs, just in virtue of their formal structure, are not sufficient for any mental phenomena to occur. For Searle, the existence of consciousness and intentionality is dependent on the causal, material microstructure of the system, not on its abstract, formal description. Consciousness can never be reduced to neurophysiology, because it is ontologically an irreducibly first-person phenomenon, believes Searle.

Patricia Churchland, in her earlier writings, has been less willing to participate in a very metaphysical tedious babble. She stressed that we obviously lack developed empirical theories about the nature of consciousness and representation: It might be that neither of these things will, in the end, turn out to be natural kinds. Of course, it is possible that the coevolution of psychology and neurobiology will ultimately preserve a macrolevel phenomenon that can be called consciousness that microlevel neurobiological theory will explain. Nevertheless, even if the concept of consciousness should not prove to be an empirically adequate category, it is reasonable, Churchland and Ramachandran argue, to regard sensory experiences as real states of the brain. Consequently, it is rational to try to discover empirically the neurobiological mechanisms underlying sensory experiences.

REFERENCES

Dennett, D. C. (1987). *The intentional stance*. Cambridge, MA: MIT Press.

Dennett, D. C. (1990). The myth of original intentionality. In K. A. Mohyeldin Said, W. H. Newton-Smith, R. Viale, & K. V. Wilkes, (Eds.), *Modelling the mind* (pp. 43–62). Oxford: Clarendon Press.

Dennett, D. C. (1991). *Consciousness explained*. Boston: Little, Brown.

Dennett, D. C., & Kinsbourne, M. (1992). Time and the observer: The where and when of consciousness in the brain (with commentary). *Behavioral and Brain Sciences, 15*, 183–247.

Quine, W. V. O. (1969). Ontological relativity. Reprinted in R. C. Hoy & L. N. Oaklander (Eds.), *Metaphysics* (pp. 514–531). Belmont, CA: Wadsworth (1991). (Page citations are to the reprint.)

Ryle, G. (1949). *The concept of mind*. London: Hutchinson.

1 Scientific Realism and the Problem of Consciousness

Ilkka Niiniluoto
University of Helsinki, Finland

Scientific realism is a philosophical trend that takes science seriously. This means that the results of scientific inquiry should be preferred to myths, religion, metaphysics, and uncritical common sense. For example, if we are interested in the problem of consciousness, we should bring together philosophers and scientists and see what they have to say about the human mind. But scientific realism makes another move in its reliance on science: Against positivist and instrumentalist views, it takes scientific theories seriously as fallible but progressive attempts to disclose the nature of reality—even beyond the observable things and events. Therefore, scientific realism has played an important role in rehabilitating the study of human consciousness within contemporary psychology and cognitive science.

COMMON SENSE VERSUS SCIENCE

According to the *common sense* view, the world consists primarily of middle-sized material objects (such as stones, plants, and animals). The behavior of these objects is governed by causal laws of nature. The highest of the animals is man, who is able to use artificial tools and languages for the purposes of work, communication, and thinking. Besides their material body, which includes the brain human beings have a mind or consciousness: They have feelings (pain, pleasure), emotions (love, hate), volitions (desires, wants), and thoughts (imagination, beliefs, memories). At least upon reflection, they are also able to recognize or become aware of their own mental life—and thereby become self-conscious.

The common sense conception is sometimes equated with the attitude that philosophers have called *naive realism*: The world is, in its essential features, just like our everyday experience tells us. This view assumes that we have direct and veridical access to the external world around us through sense perception and to our inner world through introspection. As we see stones and experience pain, it is unproblematic to assert that stones and pain exist in reality.

It is no doubt true that common sense shares quite a lot with naive realism. However, it should be emphasized that there is no unique world view of naive realism, because there is no unique way of perceiving the world. Our external and internal senses are strongly influenced by our practical interests, conceptual categories, and beliefs. We see things that are important for us, can be conceptualized within our language, and fit our expectations.[1] In some cultures, it was a part of their naive realism to see all things as animated by spirits, the sun and the planets as gods, old women as witches, and dreams as gates to another level of reality. Similarly, the language that we learn through our education teaches us to perceive and identify, around us and within us, many kinds of physical objects, artifacts, and mental events.

The common sense view of the educated members of the enlightened Western culture is already refined and enriched by many scientific discoveries. Greek philosophy and science included still many beliefs based on superstition and naive realism: The sun revolves around the earth, the planets influence our destinies, human beings think with their heart. But Plato and Aristotle taught us to see ourselves as rational human agents with reason and morality, the late medieval nominalists created the conception of a unique individual or person, Carl von Linné identified Homo sapiens as an animal species, Charles Darwin and later anthropologists showed that the origin of humanity is part of the evolution of life, the humanities and the social sciences have proven the rich variety of traditions and cultures. All these discoveries are now elements of folk psychology, that is, our contemporary common sense conception of ourselves.

Wilfrid Sellars (1963) defined the *manifest image* as "the framework in terms of which man came to be aware of himself as man-in-the-world" (p. 6). This is different from what he calls the *original image* where everything was still treated as a person. The manifest image is thus the framework of human agency where nature is depersonalized and man distinguishes himself from other things in the world by being a person. Sellars also allowed that this image is empirically refined by the methods of ordinary observation and inductive generalization—thereby it comes close to what has been previously called educated common sense. It is largely this manifest image of man that is the topic of "perennial philosophy" from Aristotle to the later Wittgenstein.

In the case of the inanimate nature, the manifest image contains primarily ordinary things of everyday observation. This uncritical common sense view has been radically challenged by the *scientific image* that uses the method of postulational theory construction: Theories of physics and chemistry reveal that, below the surface of ordinary observation, nature in fact is a complex system of microlevel particles and fields of energy. As the physicist Arthur Eddington pointed out in the 1920s, there are two different tables: the observable table as a solid, colored, wooden artifact and the scientist's table as a complex of unobservable subatomic particles.

It is characteristic to scientific realism to claim that the scientific image is in some way an improvement of the manifest image. Sellars himself argued that the manifest image is epistemically prior to the scientific image (i.e., the latter is an outgrowth of the former), but the scientific image is ontologically prior to the manifest image (i.e., theories that postulate unobservable entities correctively explain observational facts and regularities).[2] The latter thesis is in contradiction with the view, common to many empiricists and phenomenologists, that the true reality consists of the observable life-world open to our perception.[3]

A strong form of scientific realism asserts that only the scientific objects are real, and the manifest objects are merely appearances to human minds; this means that the scientific image will replace the manifest image. This line is taken by Sellars and his eliminativist followers. A weak form of scientific realism takes the manifest objects to be systems (or mereological sums) of scientific objects and thus allows them to be equally real as scientific objects, even though they could not exist without the existence of their constituents. Thus both of Eddington's tables are real. This typically means that the scientific image reinterprets and often corrects in some respect the claims of the manifest image (see, e.g., Bunge, 1979, 1981; Popper, 1972). For example, the belief of folk physics that physical space has an Euclidean structure is not radically false but rather approximately true; the scientific image that implies that the space is non-Euclidean explains why space appears to us to be Euclidean.

As the scientific method is fallible, scientific theories are always incomplete, uncertain, and corrigible. Therefore, the scientific image is not a unique theory but rather a succession of theories ($T_1, T_2, \ldots$) such that a later member of this series is an improvement of the earlier members. In the limit, Sellars urged, this sequence ($T_1, T_2, \ldots$) converges to a unique ideal "Peircean" limit theory T. Even though it makes sense to speak of such a convergence of theories (Niiniluoto, 1984), this *convergent realism* is not accepted by all scientific realists (cf. Rescher, 1987).

The construction of the scientific image started with the ancient philosophers of nature, who postulated atoms (without yet being able to test empirically this theoretical assumption). Even if today the interpretation of basic physical theories (quantum theory, theory of relativity) is still a matter of dispute and new theories abound (quarks, superstrings, black holes, and so forth), the scientific image of nature is relatively well established.

The same cannot be said about our scientific conception of human mental life. Psychological theories have not cumulated or converged to the same extent as physical theories. There is no consensus today about the meaning of terms like consciousness, mind, psyche, and self. The occurrence of activities that usually are identified in folk psychology as feeling, wanting, and believing does not even logically imply that *I* or my *consciousness* exists. Philosophers, psychologists, physiologists, and cognitive scientists give wildly different answers to questions of the following type: Does consciousness exist? What are its constitutive elements? How is it related to the physical world in general and brain in particular? Does the scientific image of man preserve the common sense view of human personhood and agency?

In this situation, the program of scientific realism recommends that better understanding of the nature of consciousness can be sought by the interaction and collaboration of empirical researchers and philosophers: The science of human mind should be taken seriously as a source of knowledge, but its results should be placed under a critical analysis.

PHILOSOPHY OF MIND AND SCIENTIFIC PSYCHOLOGY

Scientific psychology (i.e., the study of human mental life by means of the experimental method borrowed from the natural science) was born with the establishment of the first psychological laboratories in Germany in the 1870s. This new trend was not so much directed against the common sense view of man but rather against the religious and metaphysical speculation that was regularly conjoined with thinking about the human mind. Indeed, psychology was at that time still a part of philosophy and remained administratively in that position in many universities even until the 1940s.

The revolt against speculative metaphysics can be nicely illustrated by the development of scientific psychology. In the 19th century, with textbooks of psychology written in the Hegelian tradition of German *idealism*, it was quite possible for the author to assert without hesitation that the human soul is a simple nonmaterial and immortal thing or substance. Only in later editions, when experimental, physiological, and evolutionary approaches had made notable progress, did the author warn the reader that this question belongs to metaphysics rather than empirical psychology. Thus, even if idealism retained a relatively strong position in academic philosophy, as a theory of human consciousness it started to degenerate, because it was forced (e.g., in the work of Hermann Lotze, 1884) to adjust its a priori claims to the flow of new empirical evidence emerging from psychological laboratories.

Modern philosophy of mind was born in the early 17th century with René Descartes' *dualism* that takes matter and mind to be the two independently existing substances, the former having extension in space and time, the latter consisting of pure thinking. The Cartesian view of matter is mechanistic: The material world (including animals and our bodies) is like a machine following mechanistic laws, and the nonmaterial human soul is like a ghost in the machine. This type of dualism is called *interactionism*, because Descartes claimed that the human mind and body are able to causally interact with each other (via the pineal gland). Difficulties with this position led, in one direction, to *mechanistic materialism* that treated the whole of a human person (including his psyche) as a machine. Other versions of crude materialism, which were never popular among academic circles, claimed that mind is simply a special kind of matter.

Besides interactionist dualism and mechanistic materialism, the main rival to the various formulations of idealism was a Spinoza-type monistic *double-aspect* doctrine or *parallelism*, where mind and brain are two causally independent sides (inner and outer) of the same underlying neutral events. This view was held by the physiologist Theodor Fechner and the neurophysiologist John Hughlings Jackson, and it helped to establish a division of labor between the physiological study of man's central nervous system and psychology as the science of psyche.

Parallelism was also the background philosophy of Herbert Spencer's *The Principles of Psychology* (1855) that was the first attempt to explain the emergence of human consciousness on the basis of a (Lamarckian) theory of evolution (see Smith, 1982). According to Spencer, the interior states of consciousness can be built by permutation and combination from elementary units that are parallel to exterior physiological events or brain states. Animal and human consciousness is developed out of bodily life: When the nerve centers evolve and become more and more organized and intricate, feelings are awakened.

Other evolutionists were not convinced by this theory that appears to be mechanistic and circular (see Bowler, 1986; Gillespie, 1979). Charles Darwin himself reflected, from a materialist viewpoint, about the evolution of mind in his Notebooks in the 1830s (see Darwin, 1974). Charles Lyell and A. R. Wallace accepted the fact of evolution but urged that the natural evolution of man has needed help from a higher intelligence. W. K. Clifford, G. E. Romanes, and Ernst Haeckel supported the *panpsychist* view that the molecules of inorganic matter already contain a piece of "mind-stuff," which then appears in the brain as conscious experience. T. H. Huxley advocated *epiphenomenalism,* which takes the mind to be a shadow-like causal product of the brain but without its own causal powers.

It is interesting to note that the Darwinian evolutionism was also seen to presuppose that consciousness has causal powers: How could it otherwise have a favorable influence on the human adaptation to environment? This old argument was recently repeated by Karl Popper and John Eccles (1977) in their defense of mental causation from the psychical World 2 to the physical World 1 (see also, Popper, 1987). This idea can be understood in terms of dualistic interactionism (Eccles), but it can also be defended by the doctrine of emergence. An *emergent materialist* claims that the human mind is an evolutionary product of material nature, not capable of existing without a material basis (such as the brain), but still it has acquired a relatively independent status in the sense that there are causal influences in both directions between body and mind. A mirror image of this doctrine is *emergent idealism,* which takes the material body to be a "product" of mind or spirit. Variants of emergent materialism, where the evolution of mind is connected with language, communication, work, and other kinds of social action, were developed by American thinkers in the pragmatist tradition (Chauncey Wright, George Herbert Mead), and in the Marxist tradition by Friedrich Engels.

A clear distinction between three forms of materialism (and corresponding forms of idealism) was given by C. D. Broad in *The Mind and Its Place in Nature* (1925). Materialism may be *radical* (pure), *reductive,* or *emergent.* Radical materialism claims that everything is matter and only matter: Mental terms are delusive, because they do not refer to anything. This is the view of Democritus, Thomas Hobbes, J. O. de La Mettrie, and the later eliminative materialists. Reductive materialists instead think that mental phenomena can be in some way reduced to physical things and processes. Emergent materialism (with some qualifications Broad's own favorite) asserts that "mentality is an emergent characteristic of certain kinds of material complex" (Broad, 1925, p. 647).[4]

Given the perplexing variety of positions in the mind-body problem (see Fig. 1.1), many philosophers and scientists at the turn of the century followed Emil Du Bois-Reymond's agnosticist *ignoramus* doctrine: The relation of mind and matter is one of the riddles that will remain unsolvable forever (cf. McGinn, 1991).

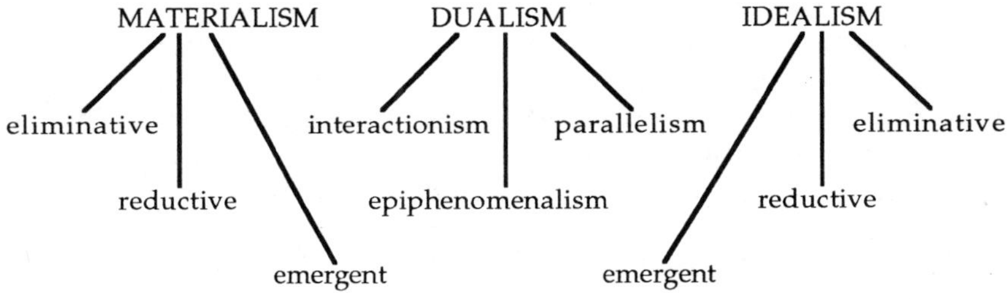

FIG. 1.1. Traditional mind–body theories

In this problem situation, the new scientific approaches to psychology appeared promising in solving at least some of the traditional questions. Among others, this was the spirit of Eino Kaila—the Finnish associate of the Vienna Circle—who was sharply critical of the endless and useless philosophical disputes in metaphysics (see Niiniluoto, Sintonen, & von Wright, 1992). Suppose we are interested in physical space and time; then we should see what the best theory of our time (viz., Einstein's theory of relativity) tells about it. Similarly, Kaila argued, to study human mind we should consult the best psychological theories of our time. Kaila claimed, against idealism and vitalism, that studies in animal and human behavior, Gestalt theory of perception, and dynamic theories of personality show our mental life to be a biological phenomenon that differs from other animals through our ability to create and use symbolic languages. The task of philosophy is to integrate this picture of man into a monistic and holistic philosophy of nature.

The program of the Vienna Circle was even more radical in its destruction of metaphysics. Rudolf Carnap argued in the 1920s that traditional philosophical theses should be reformulated in the formal mode as claims about language, not in the material mode as claims about reality—and thereby the old confusions could be overcome (see Carnap, 1967). This idea reappeared in Richard Rorty's (1980) influential thesis that the mind–body problem is an historical artifact created by Descartes—and it should be dissolved rather than positively solved. John Searle (1992) also argued that the old vocabulary employed in the analytical philosophy of mind is "obsolete" (p. 2) and "the source of our deepest philosophical difficulties" (p. 54), and therefore consciousness should be examined "on its own terms" (p. xi).

However, it seems that both ordinary language and the language of science—that is, the manifest and scientific images—are always laden with metaphysical background assumptions. We can reveal them but not get rid of them all at the same time. This has been seen clearly in the continuing philosophical debates on the interpretation of the best theories of physics and biology. The same is true in contemporary psychology and cognitive science: In spite of attempts to throw away the metaphysical past of these empirical disciplines, the old philosophical positions of Fig. 1.1 keep reappearing in the new theories.

INSTRUMENTALISM VERSUS SCIENTIFIC REALISM

Scientific realism accepts the existence of a mind-independent and knowable reality. It thus combines an ontological thesis (i.e., the existence of a mind-independent reality) with an epistemological thesis (i.e., knowledge about at least many important aspects of this reality is possible). Epistemology is tied with science: The best way of gaining knowledge about the world—including its physical, mental, and social aspects—is by theory construction. If a theory has explanatory and predictive power and passes severe tests of public observation, then there is reason to claim that its description of reality (even when it transcends the border of the observable) is true or at least truthlike. As opposed to naive realism, critical scientific realists recognize that even our best theories are often idealized, simplified, and incorrect, but nevertheless science is at least able to make progress towards the truth. (For formulations of scientific realism, see Devitt, 1991; Hempel, 1965; Leplin, 1984; Niiniluoto, 1984; Popper, 1963; Putnam, 1975, 1981, 1983; Rescher, 1987; Sellars, 1963; Smart, 1968; Tuomela, 1985.)

Many scientific realists take truth and information content to be the main epistemic utilities of theory formation, where truth is understood in the descriptive sense as the relation of correspondence between statements and reality. Scientific statements (including laws and existential claims) have truth values independent of our beliefs and interests. Other desiderata for theories include logical consistency, coherence and systematic connections with other parts of scientific knowledge, explanatory and predictive power, and independent testability by novel predictions.

Some realists take explanation to be a more basic conception than truth—or argue that "true theories and best-explaining theories coincide" (Tuomela, 1985, p. 168). For Sellars, truth is assertability relative to the ideal peirceish conceptual framework. Hilary Putnam (1981) characterized truth as ideal acceptability. Such epistemological and methodological accounts of truth have been inspired by Charles S. Peirce, whose articles founded the American pragmatism in the 1870s.

Putnam's *internal realism*—which he now distinguishes from *metaphysical realism*—combines two theses: Truth is an epistemic concept, and the world does not have a ready-made structure. The latter claim comes close to Thomas Kuhn's (1970) view that each theory or conceptual framework constructs its own ontology and thereby the concept of an "object" is theory-relative. However, Putnam (1975) himself showed how different theories may refer to the same, partly unknown objects (e.g., electrons, latent aggressions). Therefore, it can be argued that Putnam's second thesis can be combined with a nonepistemic correspondence account of truth (see Niiniluoto, 1984, 1993).

The realist interpretation of theories is opposed to *positivism* and *instrumentalism*. The latter doctrines deny that the scope of human knowledge could ever extend beyond the limits of observation, and therefore either refuse to use theoretical concepts or interpret them as merely linguistic, uninterpreted tools of observational systematization and prediction. The world-in-itself beyond the phenomena is either denied or "put in brackets"—attempts to refer to it are either false, irrelevant, or meaningless.[5]

Ernst Mach attempted to solve the psychophysical problem by the doctrine of *phenomenalism* that claims physical bodies and egos to be nothing but "complexes of sensations" (see Mach, 1959, p. 29). Carnap's *Aufbau* (1929) developed a phenomenalist constitution system that outlines (in the formal mode) a reductionist system of subjective idealism: The physical world and other minds are logical constructions from autopsychological elementary experiences (see Carnap, 1967).

The formidable technical difficulties with this program led Carnap in 1931 to join Otto Neurath's *physicalism* (see Carnap, 1959). All meaningful terms and statements are required to be translatable, via explicit definitions, to the universal and intersubjective physical language, which speaks about observable properties of observable objects. This empiricist language of physicalism is—to use terms introduced previously—a description of the things in the manifest image. For example, colors are not experienced secondary qualities but rather abilities of objects to reflect light. In the case of psychology, Carnap tried to show that all psychological sentences—about other minds, about some past or present condition of one's own mind, or general sentences—refer to physical occurrences in the body of the person in question. A typical translation of a term like *is excited* links it with the "disposition to react to certain stimuli with overt behavior of certain kinds" (Carnap, 1959, p. 197). As sentences about one's own mind are also translated to the physical language, there is no special method of introspection.

If taken as an ontological doctrine, Carnap's early physicalism is a version of reductive materialism. Methodologically it is closely associated with psychological *behaviorism,* which does not allow any terms to refer inside to the black-box of human mind. This is a typical expression of positivist operationalism in psychology. Carnap's physicalism has been called *logical behaviorism,* because it claims that every mental term is expressible, or explicitly definable, by an observational language of human behavior (Margolis, 1984). This idea was further developed by Gilbert Ryle's *The Concept of Mind* (1949), which gives an instrumentalist interpretation of the language of behavioral dispositions.

Another line was started by Carnap in 1936, when he convincingly argued that dispositional concepts are not explicitly definable by the observational language. One possibility would be then to treat such terms as nonreferring "inference tickets" (Ryle) or "intervening variables." This leads to instrumentalism. Another alternative is to liberalize the conditions of concept and theory formation by allowing the use of theoretical concepts, whose meanings are specified by the whole theoretical context but otherwise are only partially interpreted or indirectly testable by observations. Such concepts can be nevertheless understood in the realist sense as attempts to refer to really existing things. Thus, a theory need not be translatable to the observational language, because its content is not restricted to the empirical world only. With the work of Carnap, Hans Reichenbach, Carl G. Hempel, Ernest Nagel, Karl Popper, Wilfrid Sellars, and Jack Smart, this realist interpretation of theories became a major trend of analytical philosophy of science in the 1950s (see Suppe, 1977; Tuomela, 1973).

In 1965, Putnam gave a strikingly effective short answer to the question: Why theoretical terms? "*Because* without such terms we could not speak of radio stars, viruses, and elementary particles, for example—and we *wish* to speak of them, to learn more about them, to explain their behavior and properties better" (Putnam, 1975, p. 234). It is clear that this short answer should be applicable also to our wish to learn more about our own mind.

In 1956 in his "myth of Jones," Sellars explained how mental terms could be introduced to a Rylean physical-behavioral language as new theoretical terms with public criteria of use (see Sellars, 1963). In 1959 Noam Chomsky's famous criticism of B. F. Skinner's behaviorism started a new era of linguistics: The competence behind a speaker's performance was modeled on mathematical and computational terms borrowed from automata theory. A parallel development took place in the new *cognitive psychology*, where it became permissible again—in the footsteps of the earlier work by Jean Piaget and L. S. Vygotski—to construct theories that refer to real processes going on within the human consciousness (Margolis, 1984). This realist talk about beliefs and wants is also used by philosophers who endorse causalist theories of human intentional action (see Davidson, 1980; Goldman, 1970; Tuomela, 1977).

These developments suggest the possibility that good psychological theories postulate, in the scientific image of man, similar kinds of mental states and structures (consciousness, beliefs, wants, emotions, etc.) as the manifest image but provide improvements to our understanding of their nature. But also a more revolutionary break between common sense and science is possible. These two alternatives correspond to the weak and strong forms of scientific realism, respectively.

The first attempt by scientific realists to develop a materialist theory of mind was a reductionist *type-identity theory*: mental states are type-identical to brain states, just like water has turned out be identical with H_2O.[6] Unlike Carnap in the early 1930s this is a version of physicalism that allows a reference to theoretical physical entities (i.e., states of the central nervous system). Later *functionalist* theories define mental events by their causal roles or by lawlike connections to other mental and bodily states. In spite of being token-identical with brain states, mental events are not reducible to a physical language.[7] Antireductionism between psychology and physics is also characteristic to *emergent materialism*, which takes mental states to be emergent, higher-level, causally efficient properties of sufficiently complex material systems.[8] (See Fig. 1.1.) To separate such antireductionism from substance dualism, Churchland (1988) called it *property dualism*.

After being a minority view in the 1960s, *eliminative materialism* became an important research program in the 1980s thanks to Stephen Stich and Patricia and Paul Churchland (see P. Churchland, 1979, 1988, 1989; P. S. Churchland, 1986; Rosenthal, 1971; Stich, 1983). Influenced by the strong Sellarsian scientific realism, this view claims that mental terms like *belief, desire, fear, sensation, pain, joy,* and so forth belong to the radically false folk psychology and are destined to disappear (like *phlogiston* or *caloric* in physics) when the common sense framework is replaced by a matured neuroscience. In his more recent work, Paul Churchland (1989) expresssed some doubts about the realist's notion of truth and suggested moving toward some kind of pragmatism or constructivism.

Daniel Dennett's theory of mind in *Brainstorms* (1978) appears to be a sophisticated version of instrumentalism. For Dennett, a person believes that snow is white if and only if he "can be predictively attributed the belief that snow is white" (p. xvii). In ordinary cases, such mental attributions (thoughts, desires, beliefs, pains, emotions, dreams) from a third-person perspective do not pick out "good theoretical things" (p. xx). In particular, when a theory of behavior takes the "intentional stance," that is, explains and predicts the behavior of an intentional system (man or machine) by ascribing beliefs and desires to it, the claim is not that such systems really have beliefs and desires (p. 7).

John Searle (1992) advocated a realist interpretation of intrinsic mental states and events, but otherwise his position is difficult to classify because he wanted to overcome both dualism and materialism (and of course idealism). Against eliminativism, he emphasized the central importance, reality, and causal efficacy of consciousness. His *biological naturalism* takes conscious mental processes to be a causally powerful part of neurophysiological processes in the brain. In criticizing the residual dualism of strong artificial intelligence (AI), Searle (1990) claimed that "brains cause minds." Searle's (1987) recent account of consciousness is in terms of mereological supervenience: The relation of mentality to brain is "simply one more instance of the supervenience of high-order physical properties on lower-order physical properties" (p. 229). But Searle (1992) interpreted this relation in terms of bottom-up causality from neurophysiological causes to mentalistic effects (p. 125). Similarly, consciousness is a "causally emergent system feature," just like solidity and transparency: It cannot be derived from the composition of the elements and environmental relations but has to be "explained in terms of the causal interactions of the elements" (p. 111).

If consciousness has causal powers and is itself caused by the brain, this may sound like a form of interactionism—which, as we have seen, is compatible with dualism and emergent materialism. However, in *Intentionality* (1983), Searle argued that "mental states are both *caused by* the operations of the brain and *realized in* the structure of the brain" (p. 265). The obvious problem here is that the relation of causality cannot hold between two identical things. Searle tried to avoid this difficulty by interpreting his axiom "brains cause minds" as a causal relation between "different levels of description" of the same phenomena (p. 266). Such a causal influence between two levels seems to imply a dualism of properties—physical versus mental, micro versus macro, lower level versus higher level. But it also raises the suspicion that Searle's view is a form of epiphenomenalism (cf. Kim, 1989; Marras, 1993).

According to Searle, a mental event may cause a physical event: "When I raise my arm my intention in action causes my arm to go up" (Searle, 1983, p. 268). But these events are caused by and realized in neural microstructure, and because there is already a causal relation between neuron firings and physiological bodily changes, the causal relation at the macro level seems to be simply a redescription of the underlying causal process and, in this sense, superfluous (p. 270). This is confirmed by the thesis that "consciousness is causally reducible to brain processes," and the causal properties of the reduced entity are "entirely explainable in terms of the causal powers of the reducing phenomena" (Searle, 1992, pp. 114–116). But if there is no genuine agent causality or downward mental causation from the mind to the brain, then consciousness is an epiphenomenon.[9]

OBSTACLES TO REDUCTION

Human mentality has some important characteristics that every theory of consciousness has to explain—or else explain away as misconceptions. They at least give difficulties for any attempt to reduce psychology to neurophysiology—in addition to the general difficulty or impossibility of expressing necessary and sufficient conditions for types of mental states in terms of physical language.

First, our mental life has a subjective nature: It appears to be centered or united by a unique individual inner perspective (cf. Broad, 1925). Secondly, our experiences seem to have a special quality that is not captured by an external physical description (cf. Nagel, 1986). Thirdly, mental acts are intentional, directed towards objects, and thereby our thoughts have a content or semantics (cf. Fetzer, 1991).

As Searle (1992) correctly pointed out, these irreducible features are not incompatible with a scientific attitude. Following Sellars's (1963) treatment of qualia, such features can be understood as theoretical constructs that are introduced to the scientific image of man by psychological theories. If these features are real, but not preserved in the reduction to neurophysiology, the reduction is not viable. But an approach to the human mind, respecting the principles of scientific realism, is not thereby refuted.

Sentient beings have consciousness in the sense of phenomenal awareness: They feel pain, hear sounds, and so forth. But other animals differ from human beings by their lack of reflective awareness (cf. Block, 1993) or self-consciousness: They feel and hear, but they do not *know that they* feel and hear.[10] Such knowledge presupposes the ability to identify oneself as an individual extended in space and time. This in turn presupposes a conventional symbolic language that allows oneself to think about the past and the future, not only what is here and now.[11]

If subjective self-conscious awareness is an irreducible feature of our mental life, then realist theories of consciousness have to take seriously the possibility of some kind of introspection. Of course, psychological theories about introspection have to be intersubjectively testable. There is no way of going back to the Cartesian belief in the transparency of the mind: Our ability to be self-conscious about our own mind is fallible and restricted by external and internal conditions. These conditions have to be studied and specified—in the same way as psychological theories now investigate the conditions of veridical external perception.

One further item can be added to the agenda of realist theories of consciousness. Searle reminded the advocates of strong AI (or Turing machine functionalism) about the relevance of our brains to the operation of our minds: "Brains cause minds." But this should not be taken to imply that consciousness is merely a biological or neurophysiological phenomenon. It is misleading to say that mental events are "as much part of our biological natural history" as digestion and enzyme secretion (Searle, 1992, p. 1). This kind of combination of naturalism and *methodological solipsism* (Fodor, 1981), which seems to be a common denominator of many recent theories, ignores the crucially important relations we have to our material and social environment.

The phenomenologists (e.g., Dreyfus, 1972) emphasized the fact that we are beings-in-the-world through the activity of our hands and bodies. (This is slightly ironic, because phenomenology started as a program in which a pure ego introspectively reflects itself by bracketing the external world.) Such interaction not only generates most of our knowledge but can be also claimed to be constitutive of our minds. If a computer cannot have a mind until it learns to manipulate its material environment, as Marvin Minsky (1987) argued, this aspect should somehow be built into theories of human consciousness as well.

But equally important are our relations to the social and cultural enviroment. A child does not learn a language and does not grow to a human person, unless he interacts with other members of a human culture. The philosophical schools of Marxism and structuralism emphasize—even overemphasize to the extent that the subjectivity of human mind has become a problem for them—the nature of man as a cultural and social being.[12] Again the role of the environment is not only generative but also constitutive: As Tyler Burge (1986) has argued against the principle of methodological solipsism, the identification of mental states (e.g., a mother's grief for the death of her son in war or a banker's fear of the loss of his money) may presuppose reference to other social agents and institutional facts.

In Popper's terms, the human mind is at least partly a World 3 entity. Let us distinguish the physical *World 1* (physical objects and processes in nature), the subjective psychical *World 2* (mental events and processes in individual animal and human minds), and the public creations of human social activity or *World 3* (material and abstract artifacts, institutions, culture; see Fig. 1.2). Then World 1 is the one and only original reality for a radical materialist, World 2 for a subjective idealist, and World 3 for an objective idealist; a reductive materialist tries to map Worlds 2 and 3 to World 1; an emergent materialist regards Worlds 2 and 3 to be evolutionary, relatively independent real products of World 1. A dualist accepts the independent existence of Worlds 1 and 2, or Worlds 1 and 3.

Using these terms, we may say that much of the contents of the World 2 processes in individual minds is derived or learned from the cultural World 3. But the human mind is not simply a "stream of consciousness" (William James) but also a self-conscious unity. Idealistic psychology used this fact of unity as evidence for the preexistence of an immortal soul in man. But scientific evidence shows, rather, that such unity is in fact slowly developed in the first 2 years of the child's development. Therefore, it is important to note that an individual human ego or self is in fact constructed during a process, sometimes called the *psychological birth*, where a child learns the first language and also is in active interaction with the social environment.[13] According to the social theories of mind, a human self is not readymade but a social construction. In this sense, a human self is an entity in World 3.

I believe that this social dimension of human mind is even a more difficult obstacle to physicalist or neuroscientific reductions than the problems concerning the Popperian World 2: subjectivity, qualia, and intentionality. How could a neurophysiological description of my brain contain or imply all the relevant features of my mind and my natural-cultural-social environment?

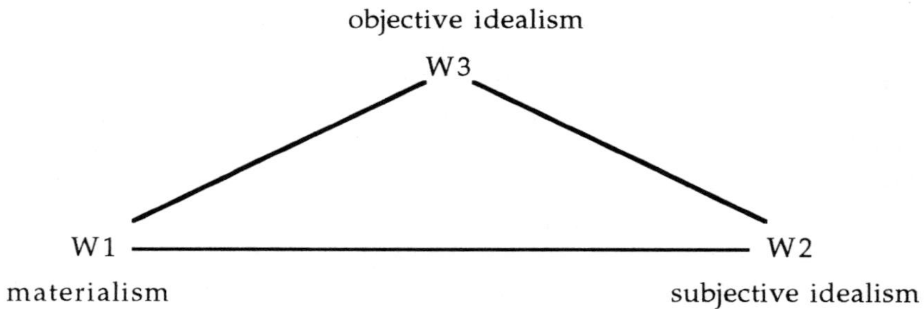

FIG. 1.2. Popper's three worlds

CONCLUDING REMARKS

Scientific realists cherish, with Charles S. Peirce, the hope that scientific theories either are true or at least converge to the truth in the long run. Peirce thought that there are real things that influence the belief formation of communities using the scientific method. Essential features of this method are the interaction of the inquirers with the object of their study and critical free discussion between the inquirers. Given these conditions, the scientific method is self-corrective and is "destined" to lead to the truth in the long run (see Peirce, 1934). Thus, if the Peircean limit theory, that is, the ultimate consensus of the scientific community, implies that consciousness exists, then consciousness exists.

Peirce's theory of truth is too optimistic, because there is no general guarantee that in real-life conditions the scientific opinion converges towards a unique limit. Even if the limit exists, in the best cases it is true only with high probability (see Niiniluoto, 1984). In spite of these reservations, it is rational to let our beliefs be guided by the most advanced results of scientific research. Sellars (1963) expressed this idea with the *scientia mensura* principle: "Science is the measure of all things" (p. 173).[14]

Whatever the strength of this scientia mensura principle is, the previous brief survey clearly indicates that we are still very far from the Peircean limit. The ontological positions of contemporary scientific realists range from eliminative materialism (P. & P. S. Churchland), reductive materialism (Smart), and emergent materialism (Popper, Bunge, Margolis) to dualism (Eccles, Swinburne) and perhaps even objective idealism (see Bohm, 1990; Shoemaker & Swinburne, 1984).

Some of them want to eliminate consciousness, others accept its existence for various reasons.

The existence of several philosophical and psychological theories about consciousness further indicates that the much-discussed pairwise contrast between folk psychology and neuroscience is misleading, because it is only a special case of the relevant issues. This comparison is inspired by the Sellarsian distinction between the manifest image and the scientific image. The radical eliminativism urges us to substitute scientific objects for the ordinary things of common sense experience; thereby the entities of folk physics and folk psychology will be eliminated. However, to be successful, the matured neuroscience of an eliminative materialist should not only replace beliefs, desires, pain, and so forth as they are ordinarily understood in everyday life, but it should eliminate also their successor concepts as they are presented in our best theories of cognitive psychology (cf. Revonsuo, 1993). It is not enough for neuroscience to supersede common sense, but it should compete with the advanced theories of cognitive psychology as well.

Given this multitude of rival theoretical approaches, it seems to me that we cannot yet apply scientia mensura to decide metaphysical controversies by scientific theories. Rather we still tend to choose our favorite type of scientific theory on the basis of our philosophical standpoint. But at the same time, with mutual cooperation between philosophy and science, we can reasonably hope to make cognitive progress by finding philosophically interesting and empirically testable psychological theories.

NOTES

[1] This Kantian theme has been elaborated in several interrelated ways in modern philosophy: Edmund Husserl called it the "intentionality of perception", Eino Kaila the "conceptuality of everyday experience", Ludwig Wittgenstein "seeing as", Thomas Kuhn the "theory-ladenness of observations". See Kaila (1979), Kuhn (1970), Hintikka (1975) and Searle (1983).

[2] This may seem similar to the traditional Aristotelean view—shared also by Galileo, Newton, and Marx—that first "resolutive" inference leads from effects to causes and then "compositive" (explanatory) inference from causes back to effects. Thereby *know that* is replaced by *know why*. However, Sellars's position is more radical, because ultimately "the common sense world of physical objects in Space and Time is unreal" (Sellars, 1963, p. 173). Note also that Sellars rejected the Lockean *hypothetico-deductive realism* that tries to causally explain our sense contents by assuming the existence of ordinary physical objects, because this would presuppose that inductive generalizations about sense contents could be formulated without using the language of physical things (Sellars, 1963, p. 87). For discussions of Sellars, see Tuomela (1977, 1985) and Pitt (1981).

[3] Phenomenalism, a radical form of empiricism, claims against common sense that an ordinary physical object is in fact a "complex of sensations." Husserl's (1970) prescientific "life-world" consists of things as we perceive them in everyday experience. Husserl argued that life-world is a "meaning fundament" for natural science, but unlike Sellars, he claimed that the mathematical theoretical science of Galileo takes us away from reality, not toward it. There is a point in this claim, because Galileo's method uses idealizations (cf. Cartwright, 1983), but it overlooks the converse process of "concretization," that is, the elimination of idealizing assumptions. Thus, a scientific realist can argue that in spite of moments of idealization theory construction as a whole is able to approach toward the truth. See Niiniluoto (1990).

[4] In fact, Broad's classification includes 27 positions, because he allowed also "neutralist" (neither materialist nor idealist) views. In my opinion, these views can be regarded as forms of objective idealism (cf. Fig. 1.2).

[5] These ideas flourished in Ernst Mach's phenomenalism, Henri Poincare's conventionalism, John Dewey's instrumentalism, and early logical empiricism (the young Carnap). Variants of these views are supported in Edmund Husserl's phenomenology, Ludwig Wittgenstein's later philosophy, the constructivist sociology of science, Bas van Fraassen's constructive empiricism, and Larry Laudan's problem-solving methodology.

[6] This view was put forward by J. Smart and D. Armstrong. See Rosenthal (1971); Feigl (1967).

[7] This view was developed by H. Putnam, D. Davidson and J. Fodor. See Haugeland (1981).

[8] See Popper and Eccles (1977); Bunge (1979, 1981); Margolis (1978); Sperry (1987); Nagel (1986). Bunge's (1981) definition of the key term *emergence* is too broad because it covers all system properties that belong to a whole but not to its parts. Hempel defined a characteristic W of an object w to be emergent relative to a theory T if W cannot be deduced or explained by means of T (from a characterization of certain kinds of parts of w with respect to certain types of attributes or relations; Hempel, 1965, p. 263. A stronger concept, which is not any more relativized to a knowledge situation, requires W to be emergent relative to *all true* theories. This concept is metaphysical because it quantifies over the unknown class of all true theories, but it is the one intended by most emergent materialists (see, however, Searle, 1992, p. 111).

[9] To be an epiphenomenon is not to be unreal: Shadows are causally produced by bodies and light rays, and they are realized in the material reality. Moreover, a shadow has causal powers—for example, someone may be frightened by my shadow. However, although a moving train is a genuine causal process, its moving shadow on the ground is merely a "pseudo-process" in Hans Reichenbach's sense. Its successive states are not caused by its earlier states, and it cannot carry information by transmitting marks (see Salmon, 1984). For an epiphenomenalist, the temporal series of mental events is only a pseudo-process.

[10] This observation throws new light on the debate on mental causation (cf. also Crane, 1992). It can be argued that mental events like occurrent desires and beliefs have causal powers through the physical event-tokens that realize them in the brain (see Marras, 1993). But *agent causality* seems to operate in a different way. Even though I feel myself hungry and see a delicious dinner in front of me, I may decide not to eat it, because I know I cannot afford it or I remember that I have promised to join my wife for dinner later in the evening. This type of mental causation presupposes that I am a self-conscious person,

capable of controlling my impulses and postponing the fulfillment of my desires. The reduction of such causality to the level of physical event-tokens seems much more problematic than the case of nonreflective mental states.

[11] Recent work in *zoosemantics* (see Sebeok, 1981) suggests that also some higher mammals (dolphins, chimpanzees) may be able to learn and use elementary symbol systems. If this is the case, it does not in any way reduce the significance of symbolic languages and self-consciousness in the human evolution.

[12] For the French structuralists (Claude Lévi-Strauss, Michel Foucault, Jacques Lacan, Louis Althusser), the human mind is like a knot in the web of cultural-social influences. The "death of an authentic subject" (Roland Barthes) became a major theme of neostructuralism (see Frank, 1989). In Lucien Sève's (1978) Marxist account, personality is "located" outside the individual. For the possibility of man as a "personal being", see Harré (1983).

[13] See Mahler, Pine, and Bergman (1975). In the Sellarsian terms, the psychological birth is connected with the creation of the manifest image in the child's development, because this is the framework "in terms of which man came to be aware of himself as man-in-the-world" (Sellars, 1963, p. 6).

[14] Tuomela (1977, 1985) formulated, following James Cornman, different versions of this principle. "Minimal" scientific realism holds that all objects do have at least those properties attributed to them by the best theories. "Moderate" scientific realism holds that physical objects are merely scientific objects, that is, possess only those properties attributed by the best theories. "Extreme" scientific realism extends this claim to all sentient objects (including persons).

REFERENCES

Block, N. (1993). Review of D. C. Dennett: Consciousness explained. *The Journal of Philosophy, 90*, 181–193.

Bohm, D. (1990). A new theory of the relationship of mind and matter. *Philosophical Psychology, 3*, 271–286.

Bowler, P. J. (1986). *Theories of human evolution*. Oxford: Blackwell.

Broad, C. D. (1925). *The mind and its place in nature*. London: Routledge & Kegan Paul.

Bunge, M. (1979). *Treatise on basic philosophy*, Vol. 4. Dordrecht: D Reidel.

Bunge, M. (1981). *Scientific materialism*. Dordrecht: D Reidel.

Burge, T. (1986). Individualism and psychology. *Philosophical Review, 45*, 3–45.

Carnap, R. (1959). Psychology in physical language. In A. J. Ayer (Ed.), *Logical positivism* (pp. 165–198). New York: The Free Press.

Carnap, R. (1967). *The logical structure of the world*. Berkeley: University of California Press.

Cartwright, N. (1983). *How the laws of physics lie*. Cambridge: Cambridge University Press.

Churchland, P. (1979). *Scientific realism and the plasticity of mind*. Cambridge: Cambridge University Press.

Churchland, P. (1988). *Matter and consciousness* (rev. ed.). Cambridge, MA: Bradford Books, MIT Press.

Churchland, P. (1989). *A neurocomputational perspective: The nature of mind and the structure of science*. Cambridge, MA: MIT Press.

Churchland, P. S. (1986). *Neurophilosophy: Toward a unified science of the mind–brain*. Cambridge, MA: MIT Press.

Crane, T. (1992). Mental causation and mental reality. *Proceedings of the Aristotelean Society, 92,* 185–202.

Darwin, C. (1974). *Metaphysics, materialism, and the evolution of mind: Early writings of Charles Darwin* (P. H. Barrett transcribor & annotator). Chicago: University of Chicago Press.

Davidson, D. (1980). *Essays on actions and events*. Oxford: Clarendon.

Dennett, D. (1978). *Brainstorms.* Montgomery: Bradford.

Devitt, M. (1991). *Realism and truth* (2nd ed.). Oxford: Blackwell.

Dreyfus, H. (1972). *What computers can't do. A critique of artificial reason.* New York: Harper & Row.

Feigl, H. (1967). *The "mental" and the "physical"*. Minneapolis: University of Minnesota Press.

Fetzer, J. (1991). *Philosophy and cognitive science.* New York: Paragon House.

Fodor, J. (1981). *Representations*. Cambridge, MA: MIT Press.

Frank, M. (1989). *What is neostructuralism?* Minneapolis: University of Minnesota Press.

Gillespie, N. C. (1979). *Charles Darwin and the problem of creation*. Chicago: University of Chicago Press.

Goldman, A. I. (1970). *A theory of human action.* Princeton: Princeton University Press.

Harré, R. (1983). *Personal being.* Oxford: Blackwell.

Haugeland, J. (Ed.). (1981). *Mind design.* Cambridge, MA: MIT Press.

Hempel, C. G. (1965). *Aspects of scientific explanation*. New York: The Free Press.

Hintikka, J. (1975). *The intentionality of intentions.* Dordrecht: D Reidel.

Husserl, E. (1970). *The crisis of European sciences and transcendental phenomenology.* Evanston: Northwestern University Press.

Kaila, E. (1979). *Reality and experience.* Dordrecht: D Reidel.

Kim, J. (1989). The myth of nonreductive materialism. *Proceedings and Addresses of the American Philosophical Association, 63,* 31–47.

Kuhn, T. S. (1970). *The structure of scientific revolutions* (2nd ed.). Chicago: University of Chicago Press.

Leplin, J. (Ed.). (1984). *Scientific realism*. Berkeley: University of California Press.

Lotze, H. (1884). *Lotze's system of philosophy* (Translated and edited by B. Bosanquet). Oxford: Oxford University Press.

Mach, E. (1959). *The analysis of sensations and the relation of the physical to the psychical.* New York: Dover.

Mahler, M. S., Pine, F., & Bergman, A. (1975). *The psychological birth of the human infant.* New York: Basic Books.

Margolis, J. (1978). *Persons and minds: The prospects of nonreductive materialism*. Dordrecht: D Reidel.

Margolis, J. (1984). *The philosophy of psychology*. Englewood Cliffs, NJ: Prentice Hall.

Marras, A. (1993). Psychophysical supervenience and nonreductive materialism. *Synthese, 95,* 275–304.

McGinn, C. (1991). *The problem of consciousness.* Oxford: Blackwell.

Minsky, M. (1987). *The society of mind.* Cambridge, MA: MIT Press.

Nagel, T. (1986). *The view from nowhere*. New York & Oxford: Oxford University Press.

Niiniluoto, I. (1984). *Is science progressive?* Dordrecht: D. Reidel.

Niiniluoto, I. (1990). Theories, approximations, idealizations. In J. Brzezinski, F. Coniglione, T. Kuipers, & L. Nowak (Eds.), *Idealization I: General problems* (pp. 9–57). *Poznan Studies in the Philosophy of the Sciences and the Humanities, 16*. Amsterdam: Rodopi.

Niiniluoto, I., Sintonen, M., & von Wright, G. H. (Eds.). (1992). *Eino Kaila and logical empiricism*. Helsinki: Acta Philosophica Fennica, 52.

Niiniluoto, I. (1993). Queries about internal realism. *Proceedings of the Beijing Conference on Realism*. Dordrecht: Kluwer.

Peirce, C. S. (1934). *Collected papers of Charles Sanders Peirce* (Vol. 5). Cambridge, MA: Harvard University Press.

Pitt, J. (1981). *Pictures, images and conceptual change: An analysis of Wilfrid Sellars' philosophy of science*. Dordrecht: D Reidel.

Popper, K. R. (1963). *Conjectures and refutations*. London: Routledge & Kegan Paul.

Popper, K. R. (1972). *Objective knowledge*. Oxford: Clarendon Press.

Popper, K. R. (1987). Natural selection and the emergence of mind. In G. Radnitzky & W. W. Bartley III (Eds.), *Evolutionary epistemology, rationality, and the sociology of knowledge* (pp. 139–155). La Salle: Open Court.

Popper, K. R., & Eccles, J. C. (1977). *The self and its brain*. Berlin: Springer International.

Putnam, H. (1975). *Mathematics, matter and method. Philosophical Papers* (Vol. 1). Cambridge: Cambridge University Press.

Putnam, H. (1981). *Reason, truth and history*. Cambridge, MA: Cambridge University Press.

Putnam, H. (1983). *Realism and reason. Philosophical papers* (Vol. 3). Cambridge: Cambridge University Press.

Rescher, N. (1987). *Scientific realism*. Dordrecht: D Reidel.

Revonsuo, A. (1993). Cognitive models of consciousness. In M. Kamppinen (Ed.), *Consciousness, cognitive schemata and relativism* (pp. 27–130). Dordrecht: Kluwer.

Rorty, R. (1980). *Philosophy and the mirror of nature*. Princeton: Princeton University Press.

Rosenthal, D. M. (Ed.). (1971). *Materialism and the mind–body problem*. Englewood Cliffs, NJ: Prentice Hall.

Ryle, G. (1949). *The concept of mind*. London: Hutchinson.

Salmon, W. (1984). *Scientific explanation and the causal structure of the world*. Princeton: Princeton University Press.

Searle, J. (1983). *Intentionality*. Cambridge, MA: Cambridge University Press.

Searle, J. (1987). Minds and brains without programs. In C. Blakemore & S. Greenfield (Eds.), *Mindwaves* (pp. 209–234). Oxford: Blackwell.

Searle, J. (1990). Is the brain's mind a computer program? *Scientific American, 262(1)*, 20–25.

Searle, J. (1992). *The rediscovery of the mind*. Cambridge, MA: MIT Press.

Sebeok, T. (1981). *The play of musement*. Bloomington: Indiana University Press.

Sellars, W. (1963). *Science, perception and reality*. London: Routledge & Kegan Paul.

Sève, L. (1978). *Man in Marxist theory and the psychology of personality*. Sussex: The Harvester Press.

Shoemaker, S., & Swinburne, R. (1984). *Personal identity*. Oxford: Blackwell.

Smart, J. J. C. (1968). *Between science and philosophy*. New York: Random House.

Smith, C. U. M. (1982). Evolution and the problem of mind: Part I, Herbert Spencer. *Journal of the History of Biology, 15*, 55–88.

Spencer, H. (1855). *The principles of psychology*. London: Longman, Brown, Green, & Longmans.

Sperry, R. W. (1987). Structure and significance of the consciousness revolutions. *The Journal of Mind and Behavior, 8(1)*, 37–65.

Stich, S. (1983). *From folk psychology to cognitive science: The case against belief*. Cambridge, MA: MIT Press.

Suppe, F. (1977). The search for philosophic understanding of scientific theories. In F. Suppe (Ed.), *The structure of scientific theories* (2nd ed.), (pp. 1–241). Urbana: University of Illinois Press.

Tuomela, R. (1973). *Theoretical concepts.* Wien & New York: Springer-Verlag.
Tuomela, R. (1977). *Human action and its explanation*. Dordrecht: D Reidel.
Tuomela, R. (1985). *Science, action and reality.* Dordrecht: D Reidel.

2 Real Consciousness

Daniel C. Dennett
Tufts University

In *Consciousness Explained* (1991a), I put forward a rather detailed empirical theory of consciousness, together with an analysis of the implications of that theory for traditional philosophical treatments of consciousness. In the critical response to the book, one of the common themes has been that my theory is not a realist theory but rather an eliminativist or verificationist denial, in one way or another, of the very reality of consciousness. Here I draw together a number of those threads, and my responses to them.[1]

It is clear that your consciousness of a stimulus is not simply a matter of its arrival at some peripheral receptor or transducer; most of the impingements on your end organs of sense never reach consciousness, and those that do become elements of your conscious experience somewhat after their arrival at your peripheries. To put it with bizarre vividness, a live, healthy eyeball disconnected from its brain is not a seat of visual consciousness—or at least that is the way we are accustomed to think of these matters. The eyes and ears (and other end organs) are entry points for raw materials of consciousness, not the sites themselves of conscious experiences. So visual consciousness must happen in between the eyeball and the mouth—to put it crudely. Where?

To bring out what the problem is, let me pose an analogy: you go to the racetrack and watch three horses, Able, Baker, and Charlie, gallop around the track. At pole 97 Able leads by a neck, at pole 98 Baker, at

[1] This paper draws on Dennett and Kinsbourne (1992b); Dennett (1993), and Dennett (in press a, in press b).

pole 99 Charlie, but then Able takes the lead again, and then Baker and Charlie run neck and neck for awhile, and then, eventually all the horses slow down to a walk and are led off to the stable. You recount all this to a friend, who asks "Who won the race?" and you say, well, because there was no finish line, there's no telling. It wasn't a real race, you see, with a finish line. First one horse led and then another, and eventually they all stopped running.

The event you witnessed was not a real race, but it was a real event—not some mere illusion or figment of your imagination. Just which kind of an event to call it is perhaps not clear, but whatever it was, it was as real as real can be.

Now consider a somewhat similar phenomenon. You are presented with some visual stimuli, A followed by B, that provoke various scattered chains of events through your brain. First the chain of events initiated by A dominates cortical activity in some regions, and then chain B takes over, and then perhaps some chain C, produced by an interaction between chain A and chain B, enjoys a brief period of dominance, to be followed by a resurgence of chain A. Then eventually all three (or more) chains of events die out, leaving behind various traces in memory. Someone asks you: Which stimulus did you become conscious of first: A or B (or C)?

If Descartes' view of consciousness were right, there would be a forthright way of telling. Descartes held that for an event to reach consciousness, it had to pass through a special gateway—which we might call the Cartesian bottleneck or turnstile—that Descartes located in the pineal gland or epiphysis. But everybody knows that Descartes was wrong. Not only is the pineal gland not the fax machine to the soul; it is not the Oval Office of the brain. It is not the "place where it all comes together" for consciousness, nor does any other place in the brain fit this description. I call this mythic place in the brain where it all comes together (and where the order of arrival determines the order of consciousness) the Cartesian theater. There is no Cartesian theater in the brain. That is a fact. Besides, if there were, what could happen there? Presumably, we care about consciousness because it matters what happens in our conscious lives—for instance, pains that never become conscious are either not pains at all, or if they are, they are not the sort that cause us suffering (such as the pains that keep us from sleeping in contorted positions and those unfortunates congenitally insensitive to pain). But if what happens in consciousness matters, surely it must make a difference—something must get done because of what happened there. However, if all the important work gets done at a point, (or just within the narrow confines of the pea-sized pineal gland) how does the rest of the brain play a role in it?

All the work that was dimly imagined to be done in the Cartesian theater has to be done somewhere, and no doubt it is distributed around in the brain. This work is largely a matter of responding to the "given" by taking it—by responding to it with one interpretive judgment or another. This corner must be turned somehow by any model of observation. On the traditional view, all the taking is deferred until the raw given, the raw materials of stimulation, have been processed in various ways. Once each bit is "finished," it can enter consciousness and be appreciated for the first time. As C. S. Sherrington (1934) put it, "The mental action lies buried in the brain, and in that part most deeply recessed from outside world that is furthest from input and output."

In the multiple drafts model developed by Marcel Kinsbourne and me (Dennett, 1991a; Dennett & Kinsbourne, 1992a), this single unified taking is broken up in cerebral space and real time. We suggest that the judgmental tasks are fragmented into many distributed moments of microtaking (Damasio, 1989; Kinsbourne, 1988). The novelty lies in how we develop the implications of this fragmentation.

It may seem at first as if we are stuck with only three alternatives:

1. Each of these distributed microtakings is an episode of unconscious judgment, and the consciousness of the taken element must be deferred to some later process (what we call the Stalinesque show trial in a Cartesian theater). But then how long must each scene wait, pending potential revision, before the curtain rises on it?
2. Each of these distributed microtakings is an episode of conscious judgment (multiple minicinemas). But then why don't we all have either a kaleidoscopic and jumbled "stream of consciousness" or (if these distributed microtakings are not "co-conscious") a case of "multiple selves"? Is our retrospective sense of unified, coherent consciousness just the artifact of an Orwellian historian's tampering with memory? As several commentators have asked, how can the manifest coherence, seriality, or unity of conscious experience be explained?
3. Some of the distributed microtakings are conscious and the rest are not. The problem then becomes: Which special property distinguishes those that are conscious, and how do we clock the onset of their activity? And, of course, because distributed microtakings may occur slightly "out of order," which "mechanism" serves to unify the scattered microtakings that are conscious, and in each case, does it

operate before or after the onset of consciousness (i.e., which phenomena are Orwellian and which are Stalinesque)?

Our view is that there is a yet a fourth alternative:

4. The creation of conscious experience is not a batch process but a continuous process. The microtakings have to interact. A microtaking, as a sort of judgment or decision, can't just be inscribed in the brain in isolation; it has to have its consequences for guiding action and modulating further microjudgments made "in its light," creating larger fragments of what we call narrative. However it is accomplished in particular cases, the interaction of microtakings has the effect that a modicum of coherence is maintained, with discrepant elements dropping out of contention and without the assistance of a Master Judge. Because there is no Master Judge, there is no further process of being-appreciated-in-consciousness, so the question of exactly when a particular element was consciously (as opposed to unconsciously) taken admits no nonarbitrary answer.

So there is a sort of stream of consciousness, with swirls and eddies, but—and this is the most important "architectural" point of our model—there is no bridge over the stream. Those of you familiar with A. A. Milne and Winnie the Pooh will appreciate the motto: "You can't play Pooh-sticks with consciousness."

As realists about consciousness, we believe that there has to be something—some property K—that distinguishes conscious events from nonconscious events. Consider the following candidate for property K: A contentful event becomes conscious if and when it becomes part of a temporarily dominant activity in cerebral cortex (Kinsbourne, 1988, and in press). This is deliberately general and undetailed, and it lacks any suggestion of a threshold. How long must participation in this dominance last and how intense or exclusive does this dominance need to be for an element to be conscious? There is no suggestion of a principled answer. Such a definition of property K meets the minimal demands of realism, but threatens the presumed distinction between Orwellian and Stalinesque revisions. Suppose some contentful element briefly flourishes in such a dominant pattern but fades before leaving a salient, reportable trace on memory. (A plausible example would be the representation of the first stimulus in a case of metacontrast—a phenomenon described in detail in Dennett, 1991a, and Dennett and Kinsbourne, 1992a.) Would this support an Orwellian or a Stalinesque model? If the element participated for "long enough," it would be "in consciousness" even if it never was properly remembered (Orwell), but if it faded "too early," it would never quite make it into the privileged category, even if it left some traces in memory (Stalin). But how long is long enough? There is no way of saying. No discontinuity divides the cases in two.

The analogy with the horse race that wasn't a horse race after all should now be manifest. Commentators on our theory generally agree with us that (a) the time of representing should not be confused with the time represented, and (b) there is no privileged place within the brain where it all comes together. They do not all agree with us, however, that it follows from (a and b) that the Orwellian/ Stalinesque distinction must break down at some scale of temporal resolution, leaving no fact of the matter about whether one is remembering misexperiences or misremembering experiences. Here, some claim, we have gone overboard, lapsing into verificationism or eliminativism or anti-realism or some other gratuitously radical position. This is curious, for we consider our position to be unproblematically realist and materialist: Conscious experiences are real events occurring in the real time and space of the brain, and hence they are clockable and locatable within the appropriate limits of precision for real phenomena of their type. (For an extended defense of this version of realism, see Dennett, 1991b.) Certain questions one might think it appropriate to ask about them, however, have no answers, because these questions presuppose inappropriate—unmotivatable—temporal and spatial boundaries that are more fine-grained than the phenomena admit.

In the same spirit we are also realists about the British Empire—it really and truly existed, in the physical space and time of this planet—but, again, we think that certain questions about the British Empire have no answers, simply because the British Empire was nothing over and above the various institutions, bureaucracies, and individuals that composed it. The question "Exactly when did the British Empire become informed of the truce in the War of 1812?" cannot be answered. The most that can be said is "sometime between December 24, 1814 and mid-January, 1815." The signing of the truce was one official, intentional act of the Empire, but the later participation by the British forces in the Battle of New Orleans was another, and it was an act performed under the assumption that no truce had been signed. Even if we can give precise times for the various moments at which various officials of the Empire became informed, no one of these moments can be singled out—except arbitrarily—as the time the Empire itself was informed. Similarly, because you are nothing over and above the various subagencies and processes in your nervous system that compose you, the following question is always a trap: "Exactly when did I (as opposed to various parts of my brain) become informed (aware, conscious) of some event?" Conscious experience, in our view, is a succession of states constituted by various processes occurring in the brain and not something over and above these processes that is caused by them.

The idea is still very compelling, however, that realism about consciousness guarantees that certain questions have answers (even if they are currently unknowable), and I have recently been fielding a variety of objections on that theme. Michael Lockwood, a commentator who is happy to declare himself to be a Cartesian materialist, nicely summed up the intuition that I am calling in question. "Consciousness," he said, with an air of enunciating something almost definitional in its status, "is the leading edge of perceptual memory" (Lockwood, 1993, p. 60).

"'Edge'?" I retorted. "What makes you think there is an edge?" Consider what must be the case if there is no such edge. The only difference between the Orwellian and Stalinesque treatment of any phenomenon is whether or not the editing or adjustment or tampering occurs before or after a presumed moment of onset of consciousness for the contents in question. The distinction can survive only if the debut into consciousness for some content is at least as accurately timable as the events of microtaking (the binding, revising, interpreting, etc.) whose order relative to the onset of consciousness defines the two positions. If the onset of consciousness is not so sharply marked, the difference between prepresentation Stalinesque revision and postpresentation Orwellian revision disappears and is restorable only by arbitrary fiat.

This is not verificationism in any contentious sense. Another analogy makes this clear. Andy Warhol anticipated a future in which each person would be famous for 15 minutes. What he nicely captured in this remark was a *reductio ad absurdum* of a certain (imaginary) concept of fame. Would that be fame? Has Warhol described a logically possible world? If we pause to think about his example more carefully than usual, we see that what makes the remark funny is that he has stretched something beyond the breaking point. It is true, no doubt, that thanks to the mass media, fame can be conferred on an anonymous citizen almost instantaneously (Rodney King comes to mind), and thanks to the fickleness of public attention, fame can evaporate almost as fast. But Warhol's rhetorical exaggeration of this fact carries us into the absurdity of Wonderland. We have yet to see an instance of someone being famous for just 15 minutes, and in fact we never will. Let some one citizen be viewed for 15 minutes by hundreds of millions of people, and then—unlike Rodney King—be utterly forgotten. To call that fame would be to misuse the term (ah yes, an ordinary language move and a good one, if used with discretion).

If that is not obvious, then let me raise the ante: Could a person be famous—not merely attended-to-by-millions-of-eyes, but famous—for 5 seconds? Every day there are in fact hundreds if not thousands of people who pass through the state of being viewed, for a few seconds, by millions of people. Consider the evening news on television, presenting a story about the approval of a new drug. Accompanying Dan Rather's voice-over, an utterly anonymous nurse is seen (by millions) plunging a hypodermic into the arm of an utterly anonymous patient. Now that's fame—right? Of course not. Being seen on television and being famous are different sorts of phenomena; the former has technologically sharp edges that the latter entirely lacks.

What I have argued, in my attack on the Cartesian theater, is that being an item in consciousness is not at all like being on television; it is, rather, a species of mental fame—almost literally. Consciousness is cerebral celebrity—nothing more and nothing less. Those contents are conscious that persevere, that monopolize resources long enough to achieve certain typical and symptomatic effects—on memory, on the control of behavior, and so forth. Not every content can be famous, for in such competitions, there must be more losers than winners. And unlike the world of sports, winning is everything. There are no higher honors to be bestowed on the winners, no Hall of Fame to be inducted into. In just the way that a Hall of Fame is a redundant formality (if you are already famous, election is superfluous, and if you are not, election probably won't make the difference), there is no induction or transduction into consciousness beyond the influence already secured by winning the competition and thereby planting lots of hooks into the ongoing operations of the brain.

Instantaneous fame is a disguised contradiction in terms, and it follows from my proposed conception of what consciousness is that an instantaneous flicker of consciousness is also an incoherent notion. Those philosophers who see me as underestimating the power of future research in neuroscience when I claim that no further discoveries from that quarter could establish that there was indeed a heretofore undreamt-of variety of evanescent—but genuine—consciousness might ask themselves if I similarly undervalue the research potential of sociology when I proclaim that it is inconceivable that sociologists could discover that Andy Warhol's prediction had come true. This could only make sense, I submit, to someone who is still covertly attached to the idea that consciousness (or fame) is the sort of semimysterious property that might be discovered to be present by the tell-tale ticking of the phenomenometer or famometer (patent pending!).[2]

Consider a question nicely parallel to the Orwell-Stalin stumpers I declare to be vacuous: Did Jack Ruby become famous before Lee Harvey Oswald died? Hmm. Well, hundreds of millions witnessed him shooting Oswald on "live" television, and certainly he subsequently became famous, but did his fame begin at the instant his gun-toting arm hove into view, while Oswald was still alive and breathing? I for one do not think there is any fact we don't already know that could conceivably shed light on this question of event ordering.

The familiar ideas die hard. It has seemed obvious to many that consciousness is—must be—rather like an inner light shining, or rather like television, or rather like a play or movie presented in the Cartesian theater. If they were right, then consciousness would have to have certain features that I deny it to have. But they are simply wrong. When I point this out I am not denying the reality of consciousness at all; I am just denying that consciousness, real consciousness, is like what they think it is like.

REFERENCES

Damasio, A. (1989). The brain binds entities and events by multiregional activation from convergence zones. *Neural Computation, 1*, 123–132.

Dennett, D. C. (1991a). *Consciousness explained.* Boston: Little, Brown.

Dennett, D. C. (1991b). Real patterns. *Journal of Philosophy, 88*, 27–51.

Dennett, D. C. (1993). Living on the edge. *Inquiry, 36*, 135–159.

[2] Owen Flanagan (1992), especially pp. 14–15 and 82–83, made the claim in detail. He supposed, for instance, that the 40-herz phase-locked oscillation championed by Singer, von der Malsburg, and others and suggested by Crick and Koch as a key mechanism in consciousness, might serve as the missing ingredient that could trump my short-sighted verificationism. "Never say never," he advised me. Well, now that neuroscience's love affair with 40-herz is beginning to cool off (because of problems that were always inherent in the idea—people were asking too much of it), let's look more closely at the prospects. If Crick and Koch are taken to be proposing a mechanism for securing cerebral celebrity—the underlying mechanism by which some contents win and others lose in competition—then they are not offering a rival view but merely specifying details I left blank in my sketch. Ignore the problems and suppose they are right (I would love to have an account of the detailed mechanisms, after all). Notice that it is logically impossible for a 40-herz oscillation mechanism to resolve temporal onset questions below the two-pulse minimum of 25 milliseconds, and any plausible competition-for-entrainment model would surely require considerably more time—moving us inexorably into a window of indeterminacy of the size I postulated: several hundred milliseconds. Alternatively, if Flanagan supposed that Crick and Koch are claiming that 40-herz entrainment causes a subsequent state-change (which could be an instantaneous transduction into some new medium, for instance), then he is simply insisting on the very concept of consciousness I am challenging—that consciousness requires entrance into a "charmed circle."

Dennett, D. C. (in press a). The message is: There is no medium. *Philosophy and Phenomenological Research.*

Dennett, D. C. (in press b). Is perception the 'leading edge' of memory? In A. Spadafora (Ed.), *Memory and Oblivion*.

Dennett, D. C., & Kinsbourne, M. (1992a). Time and the observer: The where and when of consciousness in the brain. *Behavioral and Brain Sciences, 15,* 183–200.

Dennett, D. C., & Kinsbourne, M. (1992b). Escape from the Cartesian theater (response to commentators). *Behavioral and Brain Sciences, 15,* 183–200.

Flanagan, O. (1992). *Consciousness reconsidered.* Cambridge, MA: MIT Press.

Kinsbourne, M. (1988). Integrated field theory of consciousness. In A. Marcel & E. Bisiach (Eds.), *Consciousness in contemporary science* (pp. 239–256). Oxford: Oxford University Press.

Kinsbourne, M. (in press). The Distributed brain basis of consciousness.

Lockwood, M. (1993). Dennett's mind. *Inquiry, 36,* 59–72.

Sherrington, C. (1934). *The brain and its mechanism*. Cambridge: Cambridge University Press.

3 Filling in: Why Dennett is Wrong

Patricia Smith Churchland
University of California at San Diego

Vilayanur S. Ramachandran
University of California at San Diego

INTRODUCTION

It comes as a surprise to discover that the foveal area in which one has high resolution and high acuity vision is minute; it encompasses a mere 2° of visual angle—roughly, the area of the thumbnail at arm's length. The introspective guess concerning acuity in depth likewise errs on the side of extravagance; the region of crisp, fused perception is, at arm's length, only a few centimeters deep; closer in, the area of fused perception is even narrower. The eyes make a small movement—a saccade—about every 200–300 milliseconds, sampling the scene by shifting continuously the location of the fovea. Presumably interpolation across intervals of time to yield an integrated spatio-temporal representation is a major component of what brains do. Interpolation in perception probably enables generation of an internal representation of the world that is useful in the animal's struggle for survival.

The debut demonstration of the blind spot in the visual field is comparably surprising. The standard setup requires monocular viewing of a an object offset about 13°–15° from the point of fixation (Fig. 3.1). If the object falls in the region of the blind spot of the viewing eye, the object will not be perceived. Instead, the background texture and color will be seen as uniform across the region. This is generally characterized as "filling in" of the blind spot. The existence of the perceptual blind spot is owed to the specific architecture of the retina.

As shown in Figure 3.2, each retina has a region where the optic nerve leaves the retina and hence where no transducers (rods and cones) exist. This region is the blind spot. Larger than the fovea, it is about 6° in length and about 4.5° in width.

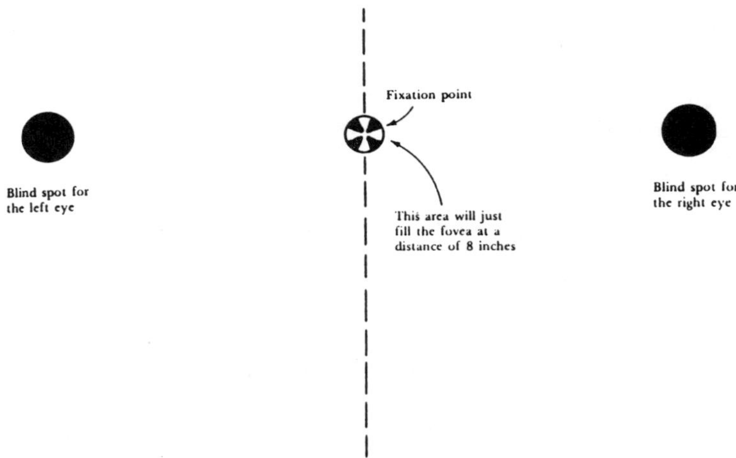

FIG. 3.1. Instructions: Close your right eye. Hold the page about 8 inches in front of you. Hold the page very straight, without tilting it. Stare at the fixation point. Adjust the angle and distance of the paper until the black spot on the left disappears. Repeat with the left eye closed. (Reproduced with permission from Lindsay & Norman, 1972.)

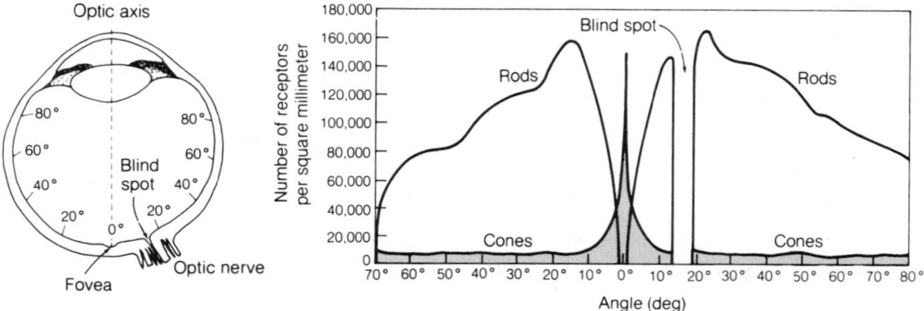

FIG. 3.2. (Left) The human eye. The blind spot (optic disk) is that region on the retina where the ganglion cells leave the retina and project to the lateral geniculate nucleus of the thalamus. (Right) The packing density of light sensitive cones is greatest at the fovea, decreasing sharply in the peripheral field. Rod density is greatest in the region that immediately surrounds the fovea and gradually decreases in the more peripheral regions. Notice that the region of the blind spot is in the peripheral field and is larger than the foveal area.

Relying on two eyes, a perceiver—even a careful and discerning perceiver—will fail to notice the blind spot, mainly because the blind regions of the two eyes do not overlap. If light from a thimble, for example, falls in the blind spot of the left eye, it will nevertheless be detected normally by the right retina, and the viewer sees a thimble. Even in the monocular condition, however, one may fail to notice the blind spot because objects whose borders extend past the boundaries of the blind spot tend to be seen as filled in, as without gaps.

What is going on when one's blind spot is seen as filled in—as without gaps in the scene? Is it analogous to acquiring the nonvisual representation (belief) that Bowser, the family dog, is under the bed, on the basis of one's visual perception of his tail sticking out? Or is it more akin to regular visual perception of the whole Bowser in one's peripheral but nonblind field? That is, is the representation itself a visual representation, involving visual experiences? In *Consciousness Explained* (1991) Dennett favored the first hypothesis that he summed up in his discussion of filling in: "The fundamental flaw in the idea of 'filling in' is that it suggests that the brain is providing something when in fact the brain is ignoring something (Dennett, 1991, p. 356).

We understand Dennett to mean that in the monocular condition, the person may represent that there is a nongappy object, say a vertical bar, in his visual field, but not because his brain generates a nongappy *visual* representation of the vertical bar. In explicating his positive view on filling in, Dennett invited us to understand filling in of the blind spot by analogy to one's impression on walking into a room wallpapered with pictures of Marilyn Monroe. "Consider how your brain must deal with wallpaper, for instance. . . . Your brain just represents *that* there are hundreds of identical Marilyns, and no matter how vivid your impression is that you see all that detail, the detail is in the world, not in your head. And no figment gets used up in rendering the seeming, for the seeming isn't rendered at all, even as a bitmap (Dennett, 1991, p. 355).

If, as instructed, we are to apply this to the case of filling in of the blind spot, presumably Dennett's point is that no matter how vivid one's impression that one sees a solid bar, one's brain actually just represents that there is a solid bar. Dennett's claim, as he clarified later, is that the brain ignores the absence of data from the region of the blind spot. In what follows, we show that contrary to Dennett, the data strongly imply that at least some instances of filling in do indeed involve the brain "providing" something.

One preliminary semantic point should be made first to forestall needless metaphysical tut-tutting. Hereafter in discussing whether someone's perception of an object, say an apple, is filled in as a convenient shorthand, we talk about whether or not "the apple is filled in." In availing ourselves of this expedient, we do *not* suppose that there might be a little (literal) apple or a (literal) picture of an apple in someone's head that is the thing that is filled in. Rather, we refer merely to some property of the brain's visual representation such that the perceiver sees a nongappy apple.

Very crudely speaking, current neurobiological data suggest that when one sees an apple, the brain is in some state that can be described as representing an apple. This representation probably consists of a pattern of activity across some set of neurons, particularly those in visual cortex, that have some specific configuration of synaptic weights and a specific profile of connectivity (P. M. Churchland, 1989). Given this general characterization of a representation, the question to address can now be rephrased: Does filling in an apple-representation consist in the visual cortex generating a representation that more closely resembles the standard case of an apple-representation of an apple in peripheral visual field? Or does it consist, as Dennett (1991) suggested, in a nonvisual representation rather like one's nonvisual representation of the dog under the bed?

Our approach to these questions assumes that a priori reflection will have value mainly as a spur to empirical investigation but not as a method that can be counted upon by itself to reveal any facts. Thought experiments are no substitute for real experiments. To understand what is going on such that the blind spot is seen as filled in (nongappy), it is important to know more about the psychological and neurobiological parameters. In addition to exploring filling in of the blind spot, other versions of visual filling in, such as the filling in experienced by subjects with cortical lesions, can also be studied. Although a more complete study would make an even wider sweep embracing modalities other than vision; for reasons of space we narrow the discussion to visual filling in.

PSYCHOPHYSICAL DATA: THE BLIND SPOT

To investigate the conditions of filling in, Ramachandran (1992) presented a variety of stimuli to subjects who were instructed to occlude one eye and fixate on a specified marker. Stimuli were then presented in various parts of the field in the region of the subject's blind spot. If a bar extends to the boundary on either side of the blind spot, but not across it, will the subject see it as complete or as having a gap? (Fig. 3.3.) Subjects see it as complete. If, however, only the lower bar segment or only the upper bar segment is presented alone, the subject does not see the bar as filled in across the blind spot (Fig. 3.4). What happens when the upper bar and the lower bar are different colors, for example, upper red, lower green? Subjects still see the bar as complete with extensions of both the red and green bar, but they do not see a border where the red and green meet, and hence they cannot say just where one color begins and the other leaves off. (For the explanation of nonperception of a border in terms of semisegregated pathways for functionally specific tasks, see Ramachandran, 1992.)

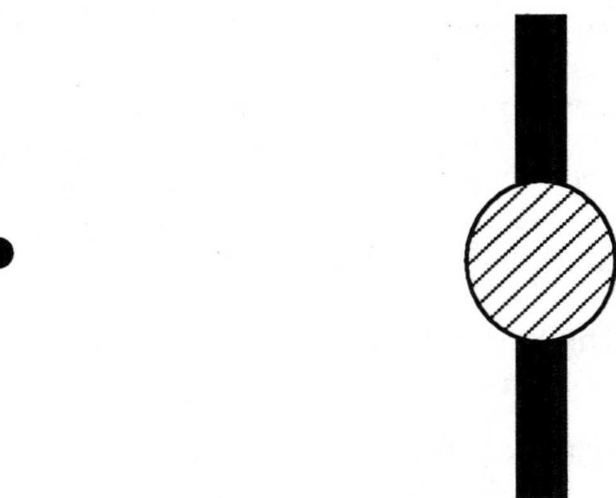

FIG. 3.3. Subjects are presented with a display consisting of two vertical bar segments separated by a gap of about 5°, and this gap is positioned to coincide with the subject's blind spot. Fixation is to the right for left eye viewing. Subjects report seeing an uninterrupted bar.

FIG. 3.4. If only the lower segment or the lower segment of the bar is presented, subjects do not complete across the blind spot.

Ramachandran also found that spokes extending to but not into the blind spot boundary were filled in, demonstrating that filling in can be very complex. Suppose there is a kind of competition between completion of a black bar across the blind spot and completion of an illusory contour lengthwise across the blind spot. Will the illusory contour or the real contour complete? Ramachandran discovered that in this test, the illusory contour typically completes (Fig. 3. 5).

Ramachandran next explored the relation between subjective completion of a figure and that figure's role in illusory motion (Fig. 3.6). The basic question is this: Does the brain treat a filled-in bar like a solid bar or like a gappy bar? In the control case, the upper gappy bar is replaced with lower gappy bar (delay about 100–200 msec). Because the gap in the upper bar is offset with respect to the gap in the lower bar, subjects see illusory motion in a diagonal direction from left to right. In the experimental (monocular) condition, the gap in the upper bar is positioned so that it falls in the subject's blind spot, and the subject sees a completed bar. Now when the upper bar is replaced by the lower bar to generate illusory motion, subjects see the bar moving vertically, nondiagonally, just as one does if a genuinely solid bar is replaced by the lower bar. This experiment shows that the brain treats a completed bar just as it treats a genuinely nongappy bar in the perception of illusory motion.

According to Dennett's characterization of filling in (1991, p. 356), the brain follows the general principle that says, in effect, "just more of the same inside the blind spot as outside." Several of Ramachandran's results are directly relevant to this claim. If filling in is just a matter of continuing the pattern outside the blind spot, then in (Fig. 3.7), subjects should see an uninterrupted string of red ovals, as a red oval fills the blank space where the blind spot is. In fact, however, subjects see an interrupted sequence; that is, they see two upper red ovals, two lower red ovals, and a white gap in between. In a different experiment, subjects are presented with a display of "bagels," with one bagel positioned so that its hole falls within the subject's blind spot (Fig. 3.8). The "more of the same" principle presumably predicts that subjects will see only bagels in the display, as one apparently sees "more Marilyns." So the blind spot should not fill in with the color of its surrounding bagel. In fact, however, this is not what happens. Subjects see bagels everywhere, save in the region of the blind spot, where they see a disk, uniformly colored.

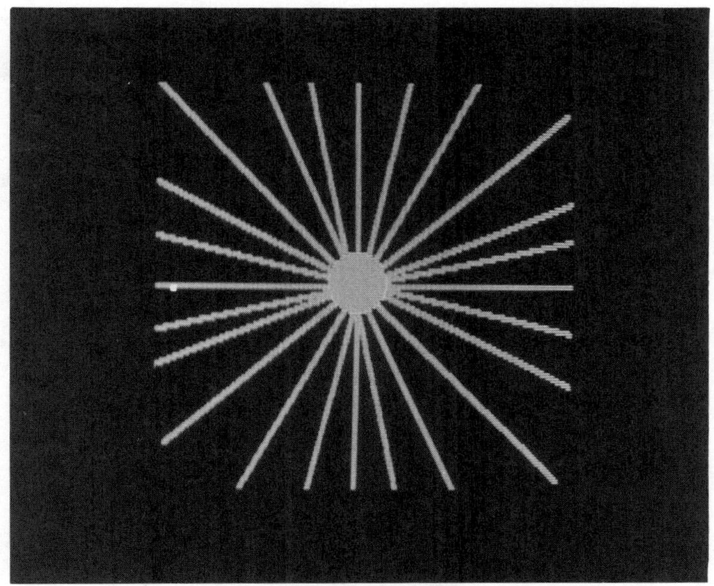

FIG. 3.5. (A) Subjects reported perceptual completion of the spokes.

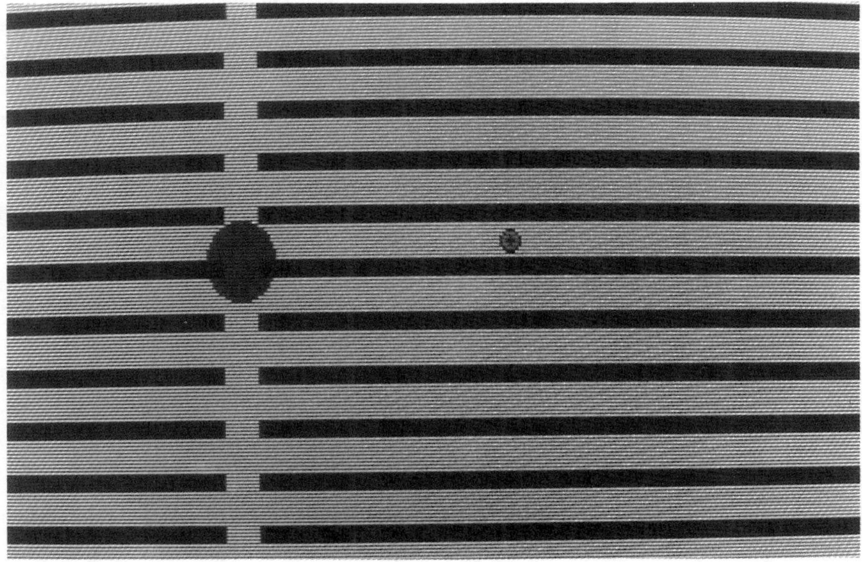

FIG. 3.5. (B) An illusory vertical strip was displayed so that a segment of the illusory contour fell on the blind spot. Subjects reported completion of the illusory strip rather than completion of the horizontal lines.

FIG. 3.6. To generate illusory motion, the upper bar is replaced by the lower bar. When the gap in the bar is located outside the blind spot, subjects see diagonal movement. When the gap coincides with the subject's blind spot, the movement appears to be vertical.

FIG. 3.7. If the display is positioned so that a circle falls in the blind spot, subjects report a gap, not completion.

FIG. 3.8. The display consists of yellow bagels and a fixation marker. The hole in one bagel (labeled *b*) coincides with the subject's blind spot. Subjects report seeing a yellow disk at this location, indicating that the yellow bagel filled in.

PSYCHOPHYSICAL DATA: CORTICAL SCOTOMATA

A lesion to early areas of visual cortex (V1, V2; i.e., areas 17, 18) typically results in a blind area in the visual field of both eyes. The standard optometric test for determining scotomata consists in flashing a point of light in various locations of the visual field. Subjects are instructed to indicate, verbally or by button pressing, when they see a flash. Using this method, the size and location of a field defect can be determined. Ramachandran, Rogers-Ramachandran, and Damasio (in press) explored the spatial and temporal characteristics of filling in of the scotoma in two patients (BM and JR).

BM had a right occipital pole lesion caused by a penetrating skull fracture. He had a paracentral left hemifield scotoma, 6° by 6°, with clear margins. JR had a right occipital lesion caused by hemorrhage and a left visual field scotoma 12° in width and 6° in height. The locations of the lesions were determined by magnetic resonance (MR) scannning. Both patients were intelligent and otherwise normal neurologically. Vision was 20/20. BM was tested 6 months and JR 8 months after the lesion events. Neither experienced his scotoma as a gap or hole in his visual field, but each was aware of the field defect. For example, each noticed some instances of false filling in of a real gap in an object. Additionally, they noticed that small, separable components of objects were sometimes unperceived and noticed as missing. For example, one subject mistook the women's room for the men's room because the *wo* of *women's* fell into the scotoma. Seeing *men's*, the subject walked directly in and quickly discovered his mistake.

In brief, the major findings of Ramachandran, Rogers-Ramachandran, and Damasio are as follows:

1. A 3° gap in a vertical line is completed across the scotoma, the completion taking about 6 seconds. The duration was determined by asking the patients to press a button when the line segment was completely filled in. Even with repeated trials, the latency remained the same.
2. One patient (JR) reported that his perception of the filled-in line segment persisted for an average of 5.3 seconds after the upper and lower lines were turned off. The delay in completion as well as the persistence of "fill" is intriguing, and it is not seen in nontraumatic blind spot filling in.
3. When the top and bottom segments of the line were misaligned horizontally by 2°, both patients first reported seeing two misaligned segments separated by a gap. After observing this for a few seconds, they spontaneously reported that the upper and lower line segments began to drift towards each other, moving into alignment, then slowly (over a period of about 10 seconds) the line segments filled in to form a single line spanning the scotoma (Fig. 3.9). The realignment and visual completion took 6.8 seconds on average.

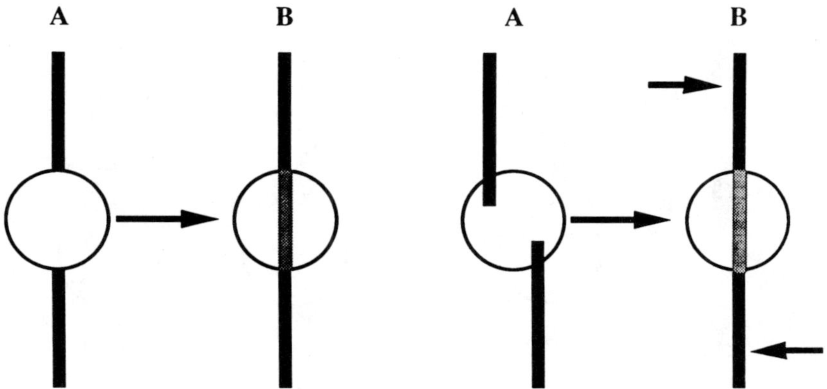

FIG. 3.9. Schematic illustration of the stimuli shown to patients. The circle represents (roughly) the region of the patient's scotoma; fixation was approximately center field. (Left) Two bar segments were displayed on either side of the scotoma. The bar was vividly completed; the process of completion took about 4–5 seconds. (Right) The vertical bar segments were misaligned in the horizontal plane. After a few seconds of viewing, patients reported the lines moving toward each other until they became colinear. They then gradually began to complete across the scotoma.

4. When viewing dynamic 2-D noise (e.g., snow on a television screen), one patient reported that the scotoma was first filled in with static (nonflickering) noise for 7 or 8 seconds before the random spots began to move and flicker. When the noise was composed of red pixels of randomly varying luminance, JR reported seeing the red color bleeding into the scotoma almost immediately, followed about 5 seconds later by the appearance of the dynamic texture.
5. When a vertical column of spots (periodicity > 2°) was used instead of a solid line, both patients clearly saw a gap. When the spacing was reduced (periodicity < 0.3°), patients reported seeing completion across the scotoma of a dotted line. These conditions were repeated using X's instead of spots, and the results were comparable (Fig. 3.10). Presenting a wavy, vertically oriented sinusoidal line (0.5 cycle/degree) with a gap matching the height of the patient's scotoma, both patients reported clearly seeing a nongappy sinusoidally wavy line.

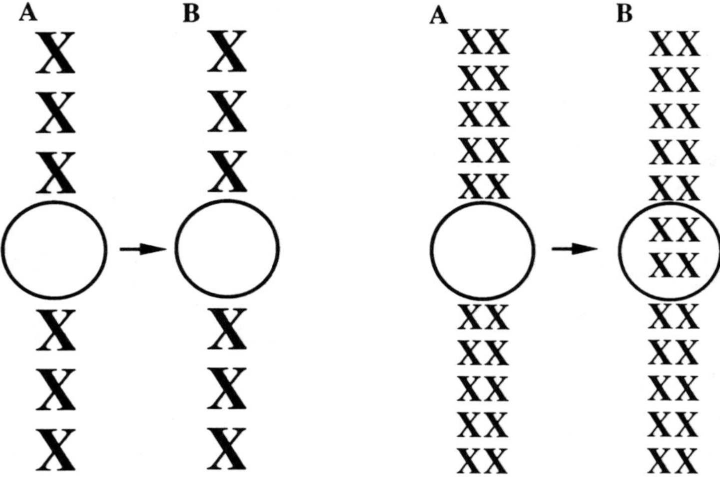

FIG. 3.10. (Left) A column of large X's was not completed across the scotoma. (Right) A column of small X's did complete. If the column consisted of small horizontal lines segments, the results were similar.

6. Each patient reported seeing illusory contours filled in across his scotoma. The experiment was similar to that performed with normal subjects (see again Fig. 3.5), save that the display was positioned so that the scotoma lined up with the gap in the stimuli. First, two horizontal line segments bordering the scotoma were presented, and as expected, they completed across the gap. Next, when an aligned array of horizontal lines were presented, the horizontal bars did not complete across the gap, and instead patients saw the vertical illusory strip complete across the scotoma.
7. Patients were presented with a checkerboard pattern, both fine (< 0.3°) and coarse (> 1.5°) grained that were readily filled in. When the checkerboard texture was subjected to counterphase flicker (7.5 Hz flicker; 0.6 check width), BM completed the flickering checks. JR, however, reported that as soon as the pattern was made to flicker, he saw nonflickering stationary checks inside his scotoma, with the result that the margins of his scotoma became entopically visible. After about 8 seconds, JR saw the dynamic checks everywhere, including his scotoma.

8. To determine whether these filling in effects might be seen in patients with laser induced paracentral retinal scotomata, the tests were repeated on two such patients. Ramachandran and colleagues found that (a) gaps in bars were not completed, (b) there was no motion or completion of misaligned bars, (c) coarse checkerboard did not complete, and (d) fine-grained 2-D random-dot textures were completed. This suggests that many of the completion effects are of cortical origin.

In the lesion studies, the time course for filling in together with the subject's reports indicate that the completion is a visual phenomenon rather than a nonvisual judgment or representation. For example, when their spontaneous reports were tested with comments such as, "You mean you *think* that the checkerboard is uniform everywhere," the patients would respond with emphatic denials like, "Doctor, I don't merely *think* it is there; I *see* that that it is there." Insofar as there is nothing in the visual stimulus corresponding to the filled in perception, it is reasonable to infer, in contrast to Dennett, that the brain is providing something, not merely ignoring something. The visual character of the phenomenon also suggests that in looking for the neurobiological mechanism, visual cortex would be a reasonable place to start.

PSYCHOPHYSICAL DATA: ARTIFICIAL SCOTOMATA

Ramachandran and Richard Gregory (1991) discovered a species of filling in readily experienced by normal subjects and conditions for which can easily be set up by anyone. The recipe is simple: Adjust the television set to snow (twinkling pattern of dots), make a fixation point with a piece of tape roughly in the middle of the screen, and place a square piece of gray paper about 1 centimeter square and roughly isoluminant to the gray of the background at a distance of about 8 centimeters from the fixation point (in peripheral vision). Both eyes may be open, and after about 10 seconds of viewing the fixation point, the square in peripheral vision vanishes completely. Thereafter, one sees a uniformly twinkling screen. Insofar as this paradigm yields filling in that is reminiscent of filling in of the blind spot and cortical scotoma, it can be described as inducing a kind of artificial blind spot. Hence Ramachandran and Gregory called it an artificial scotoma (Fig. 3.11).

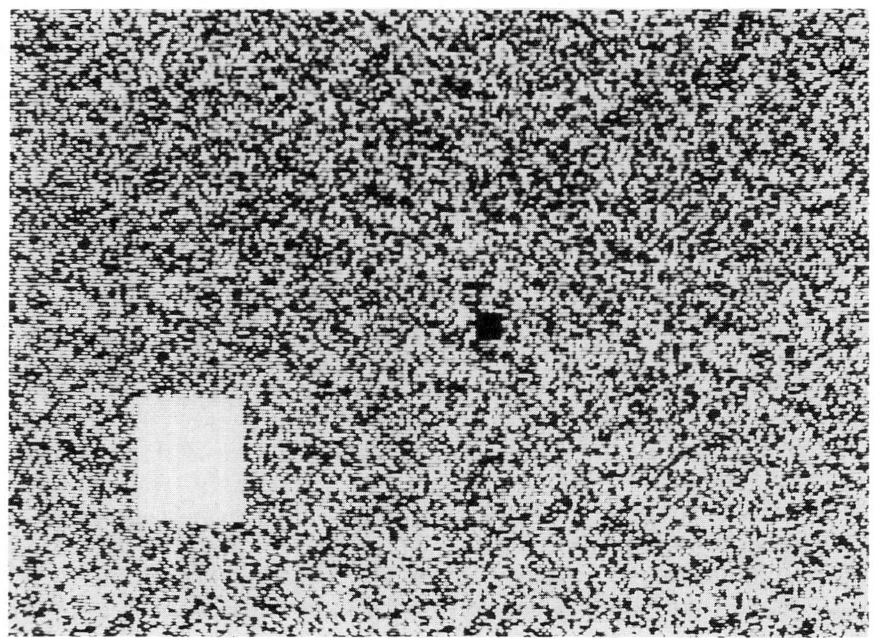

FIG. 3.11. Display for artificial scotoma conditions consists of a background texture, fixation point, and small segment in the peripheral field with a different texture, roughly isoluminant to the background.

By using a computer to generate visual displays, many different arrangements of background texture and artificial scotomata can be investigated. In exploring the variety of conditions for filling in of an artificial scotoma, Ramachandran and Gregory found a number of striking results, several of which we briefly outline as follows:

1. Subjects tended to report filling in from outside the gray square to the inside, with a time scale of about 5–10 seconds.
2. Once subjects reported filling in to be complete, the background twinkles were then turned off. Subjects now reported that they continued to see twinkling in the scotomic square for about 3–4 seconds after the background twinkles disappeared.
3. Suppose the background screen is pink, twinkles are white. The scotomic square is, as before, gray but within the square, dots are moving not randomly but coherently, left to right. Subjects report seeing a completely pink screen after about

5–10 seconds, but report that the dots in the square continue to move coherently from left to right. After a few more seconds, however, they report seeing uniformly random twinkles everywhere. Note that the artificial scotoma is in peripheral vision, where resolution is much poorer than in the foveal region (see again Fig. 3.2.) The twinkles that are filled in look just the same as the twinkles elsewhere in the peripheral field.

4. If the screen is covered in text, the peripheral square comes to be filled in with text (Fig. 3.12).

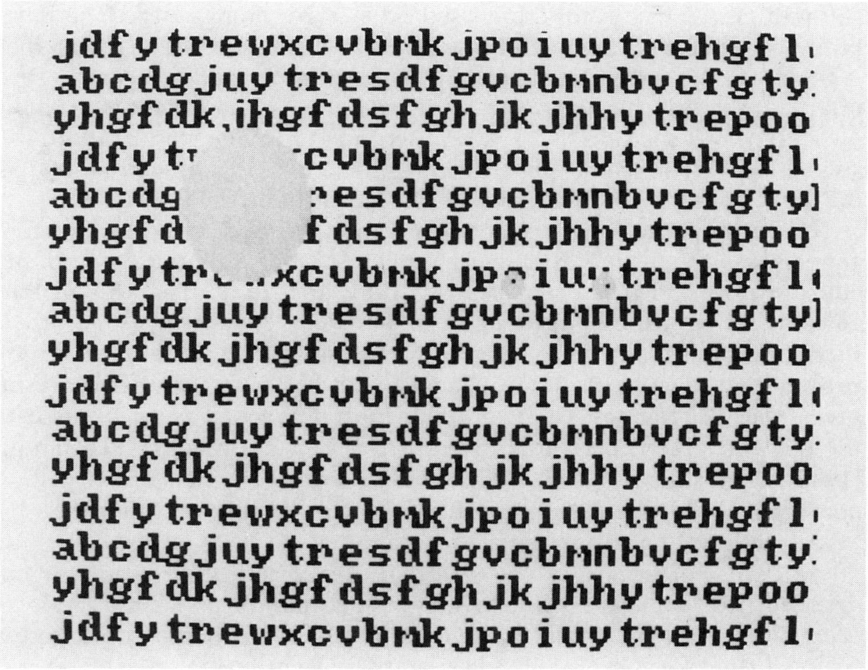

FIG. 3.12. The filling in of the artificial scotoma can be complex. In this condition, subjects report that the text fills in.

5. If a smaller square with black borders is inscribed within the region of the gray square, subjects report that the inner area does not fill in with background texture.

Many other experiments in artificial scotomata are now underway in Ramachandran's laboratory, and those few cited here mark only the first pass in exploring a salient and intriguing visual phenomenon. For our purposes here, it is perhaps enough to note that so far as we can determine, the results from artificial scotoma experiments do not confirm Dennett's hypothesis that for phenomena such as filling in, ". . we can already be *quite* sure that the medium of representation is a version of something efficient, like color-by-numbers [which gives a single label to a whole region], not roughly continuous, like bit-mapping" (Dennett, 1991, p. 354).

PSYCHOPHYSICS AND THE KRAUSKOPF EFFECT

Krauskopf (1963) discovered a remarkable filling in phenomenon. In his setup, a green disk is superimposed on a larger orange disk. The inner boundary (between green and orange) is stabilized on the retina so that it remains on exactly the same retinal location no matter how the eyes jitter and saccade, but the outer boundary moves across the retina as the eyes jitter and saccade. After a few seconds of image stabilization, the subject no longer sees a green disk; instead, the entire region is seen as uniformly orange—as filled in with the background color.

Using the Krauskopf image stabilization method to explore further aspects of the filling in phenomenon, Thomas Piantanida and his colleagues found more remarkable filling in results. It is known that adaptation to yellow light alters a subject's sensitivity to a small flickering blue light; more exactly, flicker sensitivity is reduced in the presence of a yellow adapting background. Prima facie this is odd, given that blue cones are essentially insensitive to yellow light (it is the red and green cones that are sensitive to yellow light). Piantanida (1985) asked this question: Is blue flicker sensitivity the same if yellow adaptation is obtained by subjective filling in of yellow rather than by actual yellow light illuminating the retina?

To get a perception of yellow in an area where the retina was not actually illuminated with yellow light, Piantanida presented subjects with a yellow bagel, whose inner boundary was stabilized on the retina (using a dual Purkinje eye tracker) and whose outer boundary was not stabilized. The finding was that the yellow background achieved by image stabilization was as effective in reducing blue cone flicker sensitivity as an actual yellow stimulus. This probably means therefore, that the reduction in flicker sensitivity as a function of perceived background is a cortical rather than a retinal effect. The most likely hypothesis is that cortical circumstances relevantly like those produced by retinal stimulation with yellow light are produced by yellow filling in, and hence the adaptation effects are comparable.

There is a further and quite stunning result reported by Crane and Piantanida (1983) that is especially relevant here. They presented subjects with a stimulus consisting of a green stripe adjacent to a red stripe, where the borders between them were stabilized, but the outside borders were not stabilized. After a few seconds, the colors began to fill in across the stabilized border. At this point, some observers described what they saw as a new and unnamable color that was somehow a mixture of red and green. Similar results were obtained with yellow and blue. Produced extraretinally, these visual perceptions of hitherto unperceived colors resulted from experimental manipulation of filling in mechanisms—mechanisms that actively do something, as opposed to simply ignoring something.

Dennett (1991) said of the blind spot, "The area is simply neglected" (p. 355). He said that "the brain doesn't have to 'fill in' for the blind spot since the region in which the blind spot falls is already labelled (e.g. 'plaid' or 'Marilyns' or just 'more of the same')" (p. 355). Part of the trouble with Dennett's approach to the various filling in phenomena is that he confidently prejudges what the neurobiological data at the cellular level will look like. Reasoning more like a computer engineer who knows a lot about the architectural details of the device in front of him than like a neurobiologist who realizes how much is still to be learned about the brain, Dennett jumped to conclusions about what the brain does not need to do, ought to do, and so forth.

In following two sections, we discuss neurophysiological data that conflict with Dennett's claim that "there are no homunculi, as I have put it, who are supposed to 'care about' information arising from the part of the visual field covered by the blind spot, so when nothing arrives, there is no one to complain" (Dennett, 1991, p. 357). And again, "The brain's motto for handling the blind spot could be: Ask me no questions and I'll tell you no lies" (p. 356). Although Dennett's idea may seem to have some engineering plausibility, it is really a bit of a priori neurophysiology gone wrong. Biological solutions, alas, are not easily predicted from reasonable engineering considerations. What might, from our limited vantage point, have the earmarks of sound engineering strategy, is, often as not, out of kilter with the way nature does it.

THE BLIND SPOT AND CORTICAL PHYSIOLOGY: THE GATTASS EFFECT

There are upwards of 20 cortical visual areas in each hemisphere of monkeys and probably at least that many in humans. Many of these areas are retinotopically mapped, in the sense that neighboring cells have neighboring receptive fields, that is, neighborhood points in the visual field will be represented by neighboring cells in the cortex. In particular, visual area V1 has been extensively explored (Fig. 3.13). The receptive field size of V1 cell is about 2–3° and hence is much smaller than the size of the blind spot (about 6° x 4.5°).

Ricardo Gattass and his colleagues (Fiorani, Gattass, Rosa, & Rocha-Miranda, 1990; Fiorani, Rosa, Gattass, & Rocha-Miranda, 1992; Gattass et al., 1992) were the first to try to answer the following question: How do V1 cells corresponding to the region of the blind spot for the right eye respond when the left eye is closed and a stimulus is presented to the open right eye (and vice versa)?

For ease of reference hereafter, by "Gattass condition" we denote the setup in which the experimenter records from single cells in V1 in the general area corresponding to the optic disk when the stimulus is presented to the contralateral (opposite side) eye. Call the V1 region corresponding to the the optic disk of the contralateral eye, the optic disk cortex, (ODC). The optic disk is that region of retina where no transducers exist, corresponding to that part of the visual field where the blind spot resides. Remember that if a cortical region corresponds to the optic disk for the contralateral eye, it will correspond to normal retinal area for the ipsilateral (same side) eye. (See Fig. 3.14 for projection patterns.)

The seemingly obvious answer to Gattass' question—and the answer Gattass and colleagues expected—is that the ODC cells will not respond in the monocular condition to stimuli presented in the contralateral blind spot. That is, one would predict that the cells in that region are responsive only to stimuli from the nonblind region of the ipsilateral eye. This is not what they found. Applying standard physiological mapping techniques to monkeys and using the conventional bars of light as stimuli, they tested the responses of ODC cells (left hemisphere) with the left eye closed. As they moved the bar of light around and recorded from single cells, they found that neurons in the ODC area responded very well. That is, cells corresponding to the blind spot gave consistent responses to a bar of light passing through the blind sector of the visual field. The response data did, however, show that the ODC was somewhat less neatly mapped by contralateral stimuli (i.e., in the blind spot) than by ipsilateral stimuli (i.e., in the nonblind field).

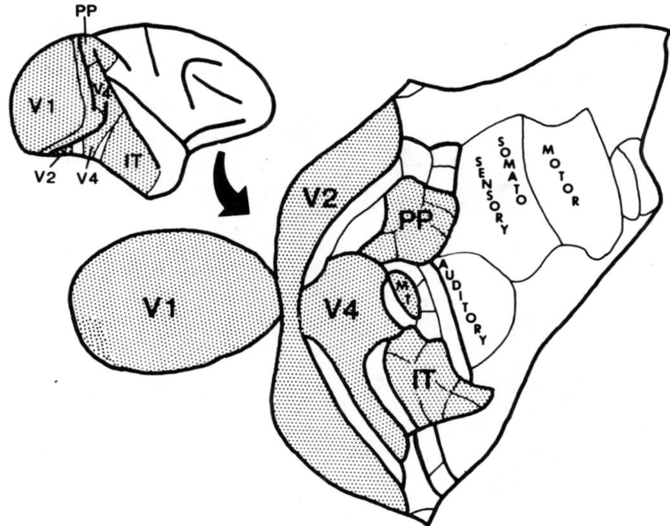

FIG. 3.13. (A) Visual areas in the cerebral cortex of the macaque, as seen in a lateral view of the right hemisphere and (arrow) in an unfolded two-dimensional map. The primary visual cortex (V1) is topographically organized. (Reproduced with permission from Van Essen & Anderson, 1990.)

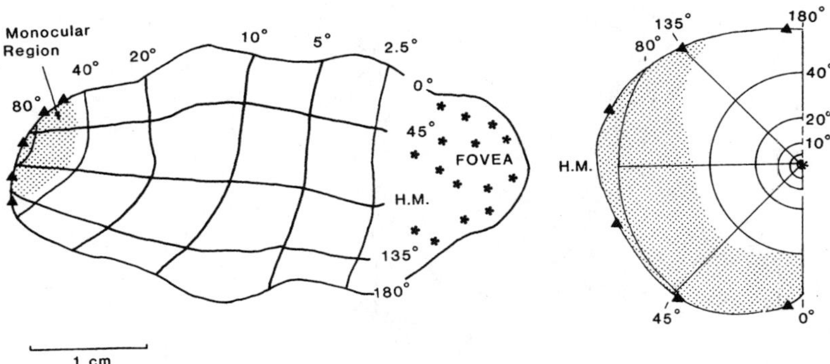

FIG. 3.13. (B) Lines of eccentricity (semicircles in the visual field drawing on lower right) map onto contours that run approximately vertically on the cortical map (lower left). Lines of constant polar angle (rays emanating from the center of gaze in the visual field) map onto contours that run approximately horizontally on the cortical map. The foveal representation (asterisks) corresponding to the central 2° radius occupies slightly more than 10% of V1. The monocular region (stipple) in the visual field occupies a very small region of the cortical map. (Reproduced with permission from Van Essen & Anderson, 1990.)

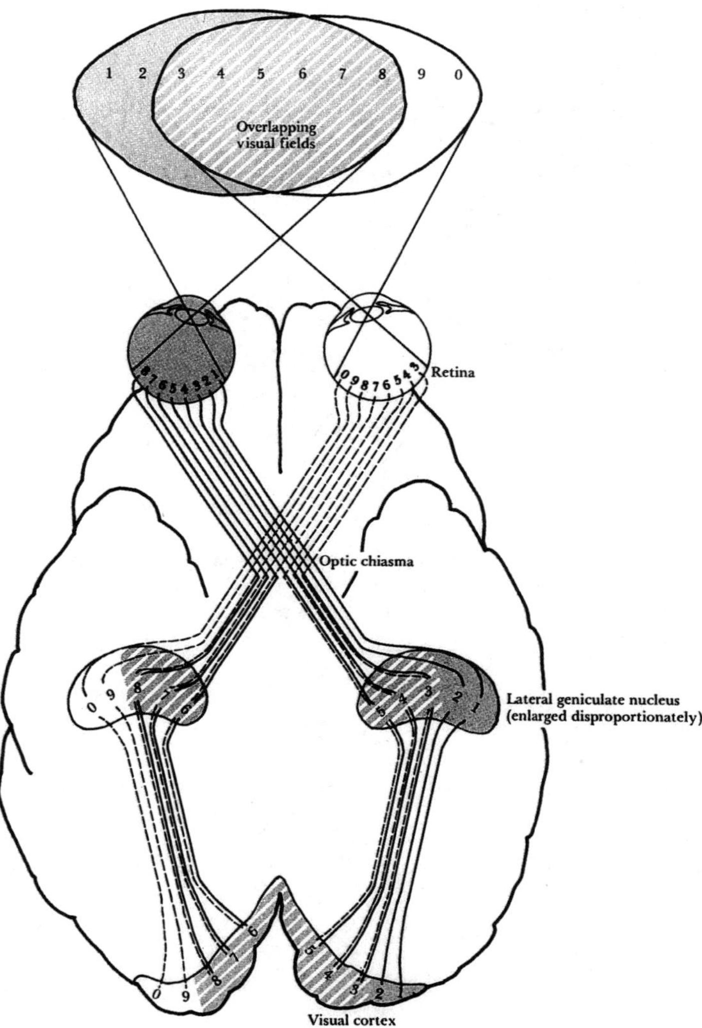

FIG. 3.14. Schematic illustration of the projection pathways from the retina to the cortex, showing which parts of the visual field are represented in specific parts of the lateral geniculate nucleus (LGN) and the visual cortex. Notice that the left hemifield projects to the right lateral geniculate nucleus (contralateral) that in turn projects to the right hemisphere. The blind spot of the left eye corresponds approximately to the region coded as "3" that is part of the central region where the fields of the two eyes overlap. By tracking 3 from the field to the retina, to the LGN and the cortex, one can track the pathway for a particular stimulus in the blind region of the left eye. (Reproduced with permission from Lindsay & Norman, 1972.)

For some cells, an excitatory receptive field—presumably an interpolated receptive field—located inside the ODC could be specified. Exploring further, they found that sweeping bars on only one end of the blind spot yielded poor responses or none at all. In other cells, they discovered that the sum of responses to two bar segments entering either end of the blind spot was comparable to the response for a single nongappy bar. This indicates that some cells in the Gattass conditions exhibit discontinuous receptive fields, presumably via interpolation signals from other neurons with neighboring receptive fields. To study the relevance of neighboring relations, Gattass and colleagues masked the area immediately surrounding the optic disk during stimulus presentations to the blind spot region of the visual field. They discovered responses of the ODC neurons were abolished in the masked condition (Figs. 3.15 and 3.16). Fifteen out of 43 neurons (mostly from layer 4ca) were found to exhibit interpolation properties across a region of the visual field at least three times the size of the classical receptive field.

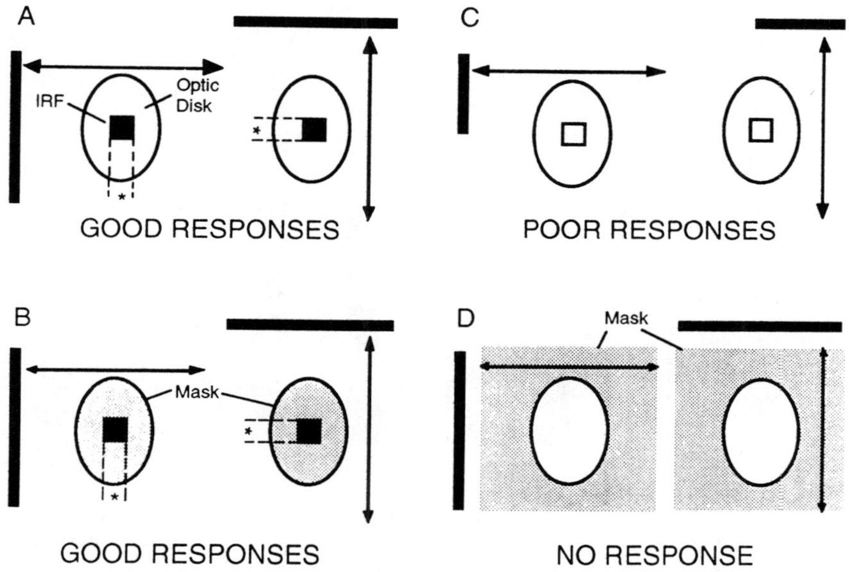

FIG. 3.15. Summary of responses of cortical neurons in the ODC region to bars and masks of varying sizes. IRF: interpolated receptive field. Asterisks indicate locations where stimulation with long bars elicited responses. (Reproduced with permission from Fiorani et al., 1992.)

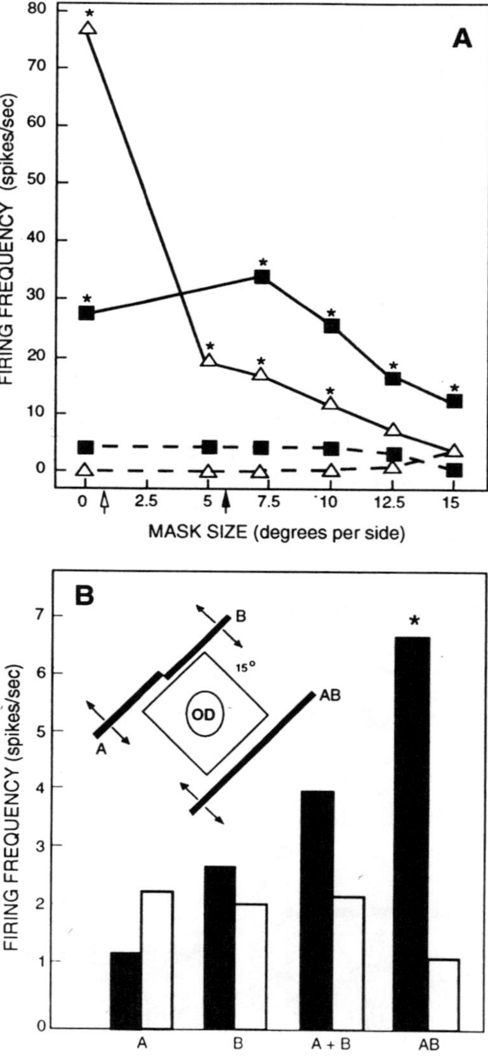

FIG. 3.16. (A) Mean response rate of a V1 cell for 10 presentations of the stimulus under different masking conditions. Triangles and filled squares linked by continuous lines show response to ipsilateral and contralateral eye, respectively, in paired trials. The lower (dotted) lines show the mean spontaneous activity where each eye is opened separately. The size of the ipsilateral classical receptive field is shown by an outlined arrow and the diameter of the OD by a filled arrow. (B) Black bars: mean response frequency of the same neuron to stimulation over a mask 15° per side. Open bars show the mean spontaneous activity in paired trials without stimulation. (Reproduced with permission from Fiorani et al., 1992.)

ARTIFICIAL SCOTOMATA AND CORTICAL PHYSIOLOGY: THE GILBERT EFFECT

How do cortical cells respond when their receptive field corresponds to the area of an artificial scotoma such as the kind Ramachandran and Gregory studied? (For ease of reference, hereafter call these cortical neurons artificial scotoma (AS) cells.) Or how do cortical cells respond when the area of both retinae from which they receive projections is lesioned? (Hereafter, call these cortical neurons retinal lesion (RL) cells.) These questions have been addressed by Charles Gilbert and colleagues of Rockefeller University. Recording from V1 in monkeys, they discovered that the receptive fields of cortical cells surrounding the cortical RL cells expanded in several minutes so that collectively they covered that part of the visual field normally covered by the RL cortical cells (Gilbert & Wiesel, 1992).

A similar result was found in the artificial scotoma experiments in cats (Pettet & Gilbert, 1991). The cortical cells in V1 surrounding the AS cortical cells very quickly expanded their receptive fields to include the area normally in the domain of the AS cells. The receptive field expansion was on the order of 3–5 fold. It was observed as soon as tests could be made (2 minutes), and it was reversible, in that once the experimental condition was removed and a standard, nonscotomic stimulus was presented, normal mapping of cortical cells was restored. Although the neurobiological basis for this modification/interpolation in receptive field properties has not yet been determined, it is conjectured that lateral interactions within the cortex are probably crucial.

The Gattass effect together with the Gilbert effect are important evidence that the receptive fields of cortical cells are dynamic and can be modified on very short time scales. What precisely this means in terms of the neurobiological mechanisms of visual experience will require many more experiments. In any case, it is unlikely that the results are irrelevant to determining whether the brain merely ignores the blind spot or whether there is an active process related to filling in. As we try to track down the neurobiology of visual awareness, the discoveries in neuroscience are important clues to the nature of visual processing and to that component of the processing relevant to visual awareness.

Do the results from Gattass and colleagues and Gilbert and colleagues mean that, contrary to Dennett's assurances, filling in *is* rendered as a bit-map? No. The choices here are not exhausted by Dennett's alternatives of "bit-map or color-by-number." We suspect that neither the bit-map metaphor nor the color-by-numbers metaphor is even remotely adequate to the kind of representation and computation in nervous systems. Indeed, the very lability of a neuron's response properties and receptive field properties means the bit-map metaphor is misleading. In order to understand more clearly how to interpret the results from Gattass and colleagues and from Gilbert and Wiesel, much more needs to be known about interpolation in neural networks and about the interaction of neurons within a mapped region and between regions. The fact is, very little is known at this point about the detailed nature of neural computation and representation, though we are at a stage where computer models highly constrained by neurobiological and psychophysical data can yield important clues (Churchland & Sejnowski, 1992).

CONCLUSION

In *Consciousness Explained,* Dennett brilliantly and quite properly debunked the idea that the brain contains a Cartesian theater wherein images and the like are displayed. But the hypothesis that filling in (perceptual completion) may sometimes involve the brain's interpolating ("contributing something rather than ignoring something") certainly needs have no truck whatever with Cartesian theaters, either implicitly or explicitly, either metaphorically or literally, either *sotto voce* or *viva voce.* Given the data from psychophysics and neurophysiology, we hypothesize that (a) the brain has mechanisms for interpolation some of which may operate early in visual processing, (b) brains sometimes visually represent completions including quite complex completions, and (c) such representation probably involves those interpolation mechanisms.

How did Dennett come to embrace a conclusion so manifestly contrary to the data, some of which was readily available when his book was published? And why does filling in play such an important role in *Consciousness Explained*? According to our analysis, the answer derives from the background behaviorist ideology that is endemic to Dennett's work from the very beginning—from his first book, *Content and Consciouness* (1969), through *Brainstorms* (1978), *Elbow Room* (1984), *The Intentional Stance* (1987), and *Conciousness Explained* (1991).

Simplified, the heart of Dennett's behaviorism is this: The conceptual framework of the mental does not denote anything real in the brain. The importance of the framework derives not from its description of neural or any other reality; rather, it is an organizing instrument that allows one to do fairly well in explaining and predicting one another's behavior, the literal unreality of qualia, and so forth, notwithstanding. How is it that the framework manages to be a useful instrument despite the unreality of its categories? According to Dennett, even though there is nothing really in the brain that corresponds to visual awareness of red, there is something or other in the brain that luckily enough, allows one to get on pretty well in making sense of people's behavior on the pretense, as it were, that the brain really does have states corresponding to awareness of red. As for filling in, Dennett's rhetorical strategy hoists it as paradigmatic of a mental thing that one mistakenly assumes to be real.

Dennett's discussions regarding the dubiousness of selected old time intuitions often fall upon receptive ears because some categories such as "the will" and "the soul" probably do not in fact correspond to anything real and because neuroscience is bound to teach many surpising things about the mental, including that some fundamental categories can be improved upon. The sweeping behaviorism-instrumentalism, however, does not follow from these observations about revisability of psychological concepts—nor even from the eliminability by cognitive neuroscience of some concepts that turn out to be the psychological counterpart of "phlogiston," "impetus," and "natural place." Thus, one may readily concur that qualia cannot be little pictures displayed in the brain's Cartesian theater and that the self is not a little person tucked away in the folds of frontal cortex. These debunking treats are, however, just the teaspoon of sugar that helps the medicine go down. And the medicine, make no mistake, is behaviorism. The elixir is Gilbert Ryle's "Ghost-Be-Gone" *The Concept of Mind* (1949). Taken regularly, it is supposed to prevent the outbreak of mental realism. Drawing on AI's conceptual repertoire of the "virtual machine," Dennett systematically argued against the neural reality and for the merely instrumental utility of mental categories generally. Dennett's engaging exposition and brilliantly inventive metaphors tend to mask the fact that this less palatable message is indeed the main message (see also McCauley, 1993).

This brief excursion through Dennett's behaviorism and instrumentalism may help explain why he is found defending assorted theses that are highly implausible from a scientific perspective: The brain does not fill in; there is nothing whatever ("no fact of the matter") to distinguish between a misperception and a misrecollection; there is no time before which one is not aware of, say, a sound, and after which one is aware; human consciousness is a virtual machine that comes into being as humans learn to talk to themselves, and so forth (Dennett, 1991).

Scientific realism, in contrast to Dennett's instrumentalism, (P. M. Churchland, 1979, 1989) proposes that scientists determine by empirical means—by converging research from experimental psychology, neuropsychology, and neuroscience—which hypotheses are probably true, and hence which categories truly apply to the mind–brain. Some categories may be largely correct, for example visual perception; some, for example memory, attention, and consciousness appear to be subdividing, budding, and regrouping; and some may be replaced by high level categories that are more empirically adequate. At this stage, it is reasonable to consider sensory experiences to be real states of the brain, states whose neurobiological properties will be discovered as cognitive neuroscience proceeds (P. M. Churchland, 1989; P. S. Churchland, 1986, 1988).

Perhaps Dennett's main achievement consists in showing the Cartesian dangers waiting to ensare those who refer to perceptual filling in by means of the expression "filling in." If so, then the achievement is primarily semantic, not empirical. Furthermore, his aim could be satisfied merely by instructing scientists on the dangers without requiring also that the very description of filling in be expunged as untrue of what goes on in the brain. In any case, one might well wonder whether Dennett overestimated the naivete amongst scientists. To judge from the literature (see references that follow), those who scientifically study perceptual completion phenomena understand perfectly well that filling in involves no Cartesian theaters, ghosts, paint, little pictures, putty knives, or homunculi. At the very least, they are no more addled by metaphor than is Dennett when he referred to the brain as "editing multiple drafts." Taken as a linguistic prohibition rather than an empirical hypothesis about the mind–brain, Dennett's thesis that the brain does not fill in sounds uncomfortably like a quirky edict of the "word-police."

ACKNOWLEDGMENTS

For helpful discussion at many stages, we thank Paul Churchland, Francis Crick, Peter Dayan, Bo Dahlbom, Ricardo Gattass, Read Montague, Diane Rogers-Ramachandran, Adina Roskies, and Oron Shagrir.

REFERENCES

Churchland, P. M. (1979). *Scientific realism and the plasticity of mind.* Cambridge: Cambridge University Press.

Churchland, P. M. (1989). *A neurocomputational perspective.* Cambridge, MA: MIT Press

Churchland, P. S. (1986). *Neurophilosophy.* Cambridge, MA: MIT Press

Churchland, P. S. (1988). Reduction and the neurobiological basis of consciousness. In A. J. Marcel & E. Bisiach (Eds.), *Consciousness in contemporary science* (pp. 273–304). Oxford: Oxford University Press.

Churchland, P. S., & Sejnowski, T. J. (1992). *The computational brain*. Cambridge, MA: MIT Press.

Crane, H. D., & Piantanida, T. P. (1983). On seeing reddish-green and yellowish-blue. *Science, 221,* 1078–1079.

Dennett, D. C. (1969). *Content and consciousness*. London: Routledge & Kegan Paul.

Dennett, D. C. (1978). *Brainstorms*. Cambridge, MA: MIT Press.

Dennett, D. C. (1984). *Elbow room*. Cambridge, MA: MIT Press.

Dennett, D. C. (1987). *The intentional stance*. Cambridge, MA: MIT Press.

Dennett, D. C. (1991). *Consciousness explained*. New York: Little, Brown.

Fiorani, M., Gattass, R., Rosa, M. G. P., & Rocha-Miranda, C. E. (1990). Changes in receptive field (RF) size of single cells of primate V1 as a correlate of perceptual completion. *Society for Neuroscience Abstracts, 16,* 1219.

Fiorani, M., Rosa, M. G. P., Gattass, R., & Rocha-Miranda, C. E. (1992). Visual responses outside the "classical" receptive field in primate striate cortex: A possible correlate of perceptual completion. *Proceedings of the National Academy of Sciences, 89,* 8547–8551.

Gattass, R., Fiorani, M., Rosa, M. G. P., Pinon, M. C. F., Sousa, A. P. B., & Soares, J. G. M. (1992). Changes in receptive field size in V1 and its relation to perceptual completion. In R. Lent (Ed.), *The visual system from genesis to maturity*. Boston: Birkhauser.

Gilbert, C. D., & Wiesel, T. N. (1992). Receptive field dynamics in adult primary visual cortex. *Nature, 356,* 150–152.

Krauskopf, J. (1963). Effect of the retinal image stabilization on the appearance of hetrochromatic targets. *Journal of the Optical Society of America, 53,* 741.

Lindsay, P. H., & Norman, D. A. (1972). *Human information processing*. New York: Academic Press.

McCauley, R. N. (1993). Why the blind can't lead the blind: Dennett on the blind spot, blind sight, and sensory qualia. *Consciousness & Cognition, 2,* 155–164.

Pettet, M. W., & Gilbert, C. D. (1991). Contextual stimuli influence receptive field size of single neurons in cat primary visual cortex. *Neuroscience Abstracts, 43,* 1–12.

Piantanida, T. P. (1985). Temporal modulation sensitivity of the blue mechanism: Measurements made with extraretinal chromatic adaptation. *Vision Research, 25,* 1439–1444.

Ramachandran, V. S. (1992). Blind spots. *Scientific American, 266,* 86–91.

Ramachandran, V. S., & Gregory, R. L. (1991). Perceptual filling in of artificially induced scotomas in human vision. *Nature, 350,* 699–702.

Ramachandran, V. S., Rogers-Ramachandran, D., & Damasio, H. (in press). Perceptual "filling in" of scotomas of cortical origin.

Ryle, G. (1949). *The concept of mind.* New York: Barnes & Noble.

Van Essen, D. C., & Anderson, C. H. (1990). Information processing in primate vision. In S. F. Zornetzer, J. L. David, & C. Lau (Eds.), *An introduction to neural and electronic networks* (pp. 43–72). San Diego: Academic Press.

4 The Problem of Consciousness[1]

John R. Searle
University of California, Berkeley

The most important scientific discovery of the present era will come when someone—or some group—discovers the answer to the following question: How exactly do neurobiological processes in the brain cause consciousness? This is the most important question facing us in the biological sciences, yet it is frequently evaded and frequently misunderstood when not evaded. In order to clear the way for an understanding of this problem, I begin to answer four questions:

1. What is consciousness?
2. What is the relation of consciousness to the brain?
3. What are some of the features that an empirical theory of consciousness should try to explain?
4. What are some common mistakes to avoid?

WHAT IS CONSCIOUSNESS?

Like most words, consciousness does not admit of a definition in terms of genus and differentia or necessary and sufficient conditions. Nonetheless, it is important to say exactly what we are talking about because the phenomenon of consciousness that we are interested in needs to be distinguished from certain other phenomena such as attention, knowledge, and self-consciousness. By consciousness, I simply mean those subjective states of sentience or awareness that begin when one

[1] This article was originally published in *Experimental and theoretical studies of consciousness*. Wiley, Chichester (Ciba Foundation Symposium 174).

goes to sleep at night or falls into a coma, or dies, or otherwise becomes, as one would say, unconscious.

Above all, consciousness is a biological phenomenon. We should think of consciousness as part of our ordinary biological history, along with digestion, growth, mitosis, and meiosis. However, though consciousness is a biological phenomenon, it has some important features that other biological phenomena do not have. The most important of these is what I have called its subjectivity. There is a sense in which each person's consciousness is private to that person, a sense in which he is related to his pains, tickles, itches, thoughts, and feelings in a way that is quite unlike the way that others are related to those pains, tickles, itches, thoughts, and feelings. This phenomenon can be described in various ways. It is sometimes described as that feature of consciousness by way of which there is something that it is like or something that it feels like to be in a certain conscious state. If somebody asks me what it feels like to give a lecture in front of a large audience, I can answer that question. But if somebody asks what it feels like to be a shingle or a stone, there is no answer to that question because shingles and stones are not conscious. The point is also put by saying that conscious states have a certain qualitative character; the states in question are sometimes described as qualia. In spite of its etymology, consciousness should not be confused with knowledge, it should not be confused with attention, and it should not be confused with self-consciousness. I consider each of these confusions in turn.

Many states of consciousness have little or nothing to do with knowledge. Conscious states of undirected anxiety or nervousness, for example, have no essential connection with knowledge.

Consciousness should not be confused with attention. Within one's field of consciousness there are certain elements that are at the focus of one's attention and certain others that are at the periphery of consciousness. It is important to emphasize this distinction because "to be conscious of" is sometimes used to mean "to pay attention to." But the sense of consciousness that we are discussing here allows for the possibility that there are many things on the periphery of one's consciousness—for example, a slight headache I now feel or the feeling of the shirt collar against my neck—that are not at the centre of one's attention. I will have more to say about the distinction between the center and the periphery of consciousness in the section titled "Some Features of Consciousness."

Finally, consciousness should not be confused with self-consciousness. There are indeed certain types of animals such as humans that are capable of extremely complicated forms of self-referential consciousness that would normally be described as self-consciousness. For example, I think conscious feelings of shame require that the agent be conscious of himself. But seeing an object or hearing a sound, for example, does not require self-consciousness. And it is not generally the case that all conscious states are also self-conscious.

WHAT ARE THE RELATIONS BETWEEN CONSCIOUSNESS AND THE BRAIN?

This question is the famous "mind–body problem." Though it has a long and sordid history in both philosophy and science, I think, in broad outline at least, it has a rather simple solution. Here it is: Conscious states are caused by lower level neurobiological processes in the brain and are themselves higher level features of the brain. The key notions here are those of *cause* and *feature*. As far as we know anything about how the world works, variable rates of neuron firings in different neuronal architectures cause all the enormous variety of our conscious life. All the stimuli we receive from the external world are converted by the nervous system into one medium, namely, variable rates of neuron firings at synapses. And equally remarkably, these variable rates of neuron firings cause all of the color and variety of our conscious life. The smell of the flower, the sound of the symphony, the thoughts of theorems in Euclidian geometry—all are caused by lower level biological processes in the brain; and as far as we know, the crucial functional elements are neurons and synapses.

Of course, like any causal hypothesis, this one is tentative. It might turn out that we have overestimated the importance of the neuron and the synapse. Perhaps the functional unit is a column or a whole array of neurons, but the crucial point I am trying to make now is that we are looking for causal relationships. The first step in the solution of the mind–body problem is: Brain processes cause conscious processes.

This leaves the question, what is the ontology, what is the form of existence, of these conscious processes? More pointedly, does the claim that there is a causal relation between brain and consciousness commit one to a dualism of physical things and mental things? The answer is a definite no. Brain processes cause consciousness, but the consciousness they cause is not some extra substance or entity. It is just a higher level feature of the whole system. The two crucial relationships between consciousness and the brain, then, can be summarized as follows: Lower level neuronal processes in the brain cause consciousness, and consciousness is simply a higher level feature of the system that is made up of the lower level neuronal elements.

There are many examples in nature where a higher level feature of a system is caused by lower level elements of that system, even though the feature is a feature of the system made up of those elements. Think of the liquidity of water or the transparency of glass or the solidity of a table, for example. Of course, like all analogies, these analogies are imperfect and inadequate in various ways. But the important thing that I am trying to get across is this: There is no metaphysical obstacle, no logical obstacle, to claiming that the relationship between brain and consciousness is one of causation and at the same time claiming that consciousness is just a feature of the brain. Lower level elements of a system can cause higher level features of that system, even though those features are features of a system made up of the lower level elements. Notice, for example, that just as one cannot reach into a glass of water and pick out a molecule and say, "This one is wet", so one cannot point to a single synapse or neuron in the brain and say, "This one is thinking about my grandmother." As far as we know anything about it, thoughts about grandmothers occur at a much higher level than that of the single neuron or synapse, just as liquidity occurs at a much higher level than that of single molecules.

Of all the theses that I am advancing in this chapter, this one arouses the most opposition. I am puzzled as to why there should be so much opposition, so I want to clarify a bit further what the issues are. First, I want to argue that we simply know as a matter of fact that brain processes cause conscious states. We do not know the details about how it works, and it may well be a long time before we understand the details involved. Furthermore, it seems to me that an understanding of how exactly brain processes cause conscious states may require a revolution in neurobiology. Given our present explanatory apparatus, it is not at all obvious how, within that apparatus, we can account for the causal character of the relation between neuron firings and conscious states. But, at present, from the fact that we do not know *how* it occurs, it does not follow that we do not know *that* it occurs. Many people who object to my solution (or dissolution) of the mind–body problem, object on the grounds that we have no idea how neurobiological processes could cause conscious phenomena. But that does not seem to me a conceptual or logical problem. That is an empirical/theoretical issue for the biological sciences. The problem is to figure out exactly how the system works to produce consciousness, and because we know that in fact it does produce consciousness, we have good reason to suppose that there are specific neurobiological mechanisms by way of which it works.

There are certain philosophical moods we sometimes get into when it seems absolutely astounding that consciousness could be produced by electrobiochemical processes, and it seems almost impossible that we would ever be able to explain it in neurobiological terms. Whenever we get in such moods, however, it is important to remind ourselves that similar mysteries have occurred before in science. A century ago it seemed extremely mysterious, puzzling, and to some people metaphysically impossible that life should be accounted for in terms of mechanical, biological, and chemical processes. But now we know that we can give such an account, and the problem of how life arises from biochemistry has been solved to the point that we find it difficult to recover, difficult to understand why it seemed such an impossibility at one time. Earlier still, electromagnetism seemed mysterious. On a Newtonian conception of the universe, there seemed to be no place for the phenomenon of electromagnetism. But with the development of the theory of electromagnetism, the metaphysical worry dissolved. I believe that we are having a similar problem about consciousness now. But once we recognize the fact that conscious states are caused by neurobiological processes, we automatically convert the issue into one for theoretical scientific investigation. We have removed it from the realm of philosophical or metaphysical impossibility.

SOME FEATURES OF CONSCIOUSNESS

The next step in the discussion is to list some (not all) of the essential features of consciousness that an empirical theory of the brain should be able to explain.

Subjectivity

As I mentioned earlier, this is the most important feature. A theory of consciousness needs to explain how a set of neurobiological processes can cause a system to be in a subjective state of sentience or awareness. This phenomenon is unlike anything else in biology, and in a sense it is one of the most amazing features of nature. We resist accepting subjectivity as a ground floor, irreducible phenomenon of nature because, since the 17th century, we have come to believe that science must be objective. But this involves a pun on the notion of objectivity. We are confusing the epistemic objectivity of scientific investigation with the ontological objectivity of the typical subject matter in science in disciplines such as physics and chemistry. Because science aims at objectivity in the epistemic sense that we seek truths that are not dependent on the particular point of view of this or that investigator, it has been tempting to conclude that the reality investigated by science must be objective in the sense of existing independently of the experiences in the human individual. But this last feature, ontological objectivity, is not an essential trait of science. If science is supposed to give an account of how the world works and if subjective states of consciousness are part of the world, then we should seek an (epistemically) objective account of an (ontologically) subjective reality, the reality of subjective states of consciousness. What I am arguing here is that we can have an epistemically objective science of a domain that is ontologically subjective.

Unity

It is important to recognize that in nonpathological forms of consciousness, we never just have, for example, a pain in the elbow, a feeling of warmth, or an experience of seeing something red, but we have them all occurring simultaneously as part of one unified conscious experience. Kant called this feature "the transcendental unity of apperception." Recently, in neurobiology it has been called "the binding problem." There are at least two aspects to this unity that require special mention. First, at any given instant all of our experiences are unified into a single conscious field. Second, the organization of our consciousness extends over more than simple instants. So, for example, if I begin speaking a sentence, I have to maintain, in some sense at least, an iconic memory of the beginning of the sentence so that I know what I am saying by the time I get to the end of the sentence.

Intentionality

Intentionality is the name that philosophers and psychologists give to that feature of many of our mental states by which they are directed at or about states of affairs in the world. If I have a belief or a desire or a fear, there must always be some content to my belief, desire, or fear. It must be about something even if the something it is about does not exist or is a hallucination. Even in cases when I am radically mistaken, there must be some mental content that purports to make reference to the world. Not all conscious states have intentionality in this sense. For example, there are states of anxiety or depression where one is not anxious or depressed about anything in particular but just is in a bad mood. That is not an intentional state. But if one is depressed about a forthcoming event, that is an intentional state because it is directed at something beyond itself.

There is a conceptual connection between consciousness and intentionality in the following respect. Though many, indeed most, of our intentional states at any given point are unconscious, nonetheless, in order for an unconscious intentional state to be genuinely an intentional state it must be accessible in principle to consciousness. It must be the sort of thing that could be conscious even if it, in fact, is blocked by repression, brain lesion, or sheer forgetfulness.

The Distinction Between the Center and the Periphery of Consciousness

At any given moment of nonpathological consciousness, I have what might be called a field of consciousness. Within that field I normally pay attention to some things and not to others. So, for example, right now I am paying attention to the problem of describing consciousness but very little attention to the feeling of the shirt on my back or the tightness of my shoes. It is sometimes said that I am unconscious of these. But that is a mistake. The proof that they are a part of my conscious field is that I can at any moment shift my attention to them. But in order for me to shift my attention to them, there must be something there that I was previously not paying attention to which I am now paying attention to.

The Gestalt Structure of Conscious Experience

Within the field of consciousness, our experiences are characteristically structured in a way that goes beyond the structure of the actual stimulus. This was one of the most profound discoveries of the Gestalt psychologists. It is most obvious in the case of vision, but the phenomenon is quite general and extends beyond vision. For example, the sketchy lines drawn in Fig. 4.1 do not physically resemble a human face.

FIG. 4.1.

If we actually saw someone on the street that looked like that, we would be inclined to call an ambulance. The disposition of the brain to structure degenerate stimuli into certain structured forms is so powerful that we will naturally tend to see this as a human face. Furthermore, not only do we have our conscious experiences in certain structures, but we tend also to have them as figures against backgrounds. Again, this is most obvious in the case of vision. Thus, when I look at the figure I see it against the background of the page. I see the page against the background of the table. I see the table against the background of the floor, and I see the floor against the background of the room, until we eventually reach the horizon of my visual consciousness.

The Aspect of Familiarity

It is a characteristic feature of nonpathological states of consciousness that they come to us with what I call the "aspect of familiarity." In order for me to see the objects in front of me as, for example, houses, chairs, people, tables, I have to have a prior possession of the categories of houses, chairs, people, tables. But that means that I assimilate my experiences into a set of categories that are more or less familiar to me. When I am in an extremely strange environment, in a jungle village, for example, and the houses, people, and foliage look exotic to me, I still perceive that as a house, that as a person, that as clothing, that as a tree or a bush. The aspect of familiarity is thus a scalar phenomenon. There can be greater or lesser degrees of familiarity. But it is important to see that nonpathological forms of consciousness come to us under the aspect of familiarity. Again, one way to consider this is to look at the pathological cases. In Capgras' syndrome, the patients are unable to acknowledge familiar people in their environment as the people they actually are. They think the spouse is not really their spouse but is an imposter, and so forth. This is a case of a breakdown in one aspect of familiarity. In nonpathological cases, it is extremely difficult to break with the aspect of familiarity. Surrealist painters try to do it. But even in the surrealist painting, the three-headed woman is still a woman, and the drooping watch is still a watch.

Mood

Part of every normal conscious experience is the mood that pervades the experience. It need not be a mood that has a particular name to it, like depression or elation; but there is always what one might call a flavor or tone to any normal set of conscious states. So, for example, at present I am not especially depressed and I am not especially ecstatic, nor indeed, am I what one would call simply blah. Nonetheless, there is a certain mood to my present experiences. Mood is probably more easily explainable in biochemical terms than several of the features I have mentioned. We may be able to control, for example, pathological forms of depression by mood-altering drugs.

Boundary Conditions

All of my nonpathological states of consciousness come to me with a certain sense of what one might call their "situatedness." Though I am not thinking about it and though it is not part of the field of my consciousness, I nonetheless know what year it is, what place I am in, what time of day it is, the season of the year it is, and usually even what month it is. All of these are the boundary conditions or the situatedness of nonpathological conscious states. Again, one can become aware of the pervasiveness of this phenomenon when it is absent. So, for example, as one gets older there is a certain feeling of vertigo that comes over one when one loses a sense of what time of year it is or what month it is. The point I am making now is that conscious states are situated and they are experienced as situated even though the details of the situation need not be part of the content of the conscious states.

SOME COMMON MISTAKES ABOUT CONSCIOUSNESS

I would like to think that everything I have said so far is just a form of common sense. However, I have to report, from the battlefronts as it were, that the approach I am advocating to the study of consciousness is by no means universally accepted in cognitive science nor even neurobiology. Indeed, until quite recently many workers in cognitive science and neurobiology regarded the study of consciousness as somehow out of bounds for their disciplines. They thought that it was beyond the reach of science to explain why warm things feel warm to us or why red things look red to us. I think, on the contrary, that it is precisely the task of neurobiology to explain these and other questions about consciousness. Why would anyone think otherwise? Well, there are complex historical reasons going back at least to the 17th century, why people thought that consciousness was not part of the material world. A kind of residual dualism prevented people from treating consciousness as a biological phenomenon like any other. However, I am not now going to attempt to trace this history. Instead, I am going to point out some common mistakes that occur when people refuse to address consciousness on its own terms.

The characteristic mistake in the study of consciousness is to ignore its essential subjectivity and to try to treat it as if it were an objective third person phenomenon. Instead of recognizing that consciousness is essentially a subjective, qualitative phenomenon, many people mistakenly suppose that its essence is that of a control mechanism or a certain kind of set of dispositions to behavior or a computer program. The two most common mistakes about consciousness are to suppose that it can be analyzed behavioristically or computationally. The Turing test disposes one to make precisely these two mistakes, the mistake of behaviorism and the mistake of computationalism. It leads one to suppose that for a system to be conscious, it is both necessary and sufficient that it has the right computer program or set of programs with the right inputs and outputs. I think one has only to state this position clearly to see that it must be mistaken. A traditional objection to behaviorism was that behaviorism could not be right because a system could behave as if it were conscious without actually being conscious. There is no logical connection, no necessary connection between inner, subjective, qualitative mental states and external, publicly observable behavior. Of course, in actual fact, conscious states characteristically cause behavior. But the behavior that they cause has to be distinguished from the states themselves. The same mistake is repeated by computational accounts of consciousness. Just as behavior by itself is not sufficient for consciousness, so computational models of consciousness are not sufficient by themselves for consciousness. The computational model of consciousness stands to consciousness in the same way the computational model of anything stands to the domain being modeled. Nobody supposes that the computational model of rainstorms in London will leave everyone all wet. But they make the mistake of supposing that the computational model of consciousness is somehow conscious. It is the same mistake in both cases.

There is a simple demonstration that the computational model of consciousness is not sufficient for consciousness. I have given it many times before so I will not dwell on it here. Its point is simply this: Computation is defined syntactically. It is defined in terms of the manipulation of symbols. But the syntax by itself can never be sufficient for the sort of contents that characteristically go with conscious thoughts. Just having zeros and ones by themselves is insufficient to guarantee mental content, conscious or unconscious. This argument is sometimes called "the Chinese room argument" because I originally illustrated the point with the example of the person who goes through the computational steps for answering questions in Chinese but does not thereby acquire any understanding of Chinese (Searle, 1980). The point of the parable is clear but it is usually neglected. Syntax by itself is not sufficient for semantic content. In all of the attacks on the Chinese room argument, I have never seen anyone come out baldly and say they think that syntax is sufficient for semantic content.

However, I now have to say that I was conceding too much in my earlier statements of this argument. I was conceding that the computational theory of the mind was at least false. But it now seems to me that it does not reach the level of falsity because it does not have a clear sense. Here is why.

The natural sciences describe features of reality that are intrinsic to the world as it exists independently of any observers. Thus, gravitational attraction, photosynthesis, and electromagnetism are all subjects of the natural sciences because they describe intrinsic features of reality. But such features such as being a bathtub, being a nice day for a picnic, being a $5 bill or being a chair, are not subjects of the natural sciences because they are not intrinsic features of reality. All the phenomena I named—bathtubs, and so forth—are physical objects and as physical objects have features that are intrinsic to reality. But the feature of being a bathtub or a $5 bill exists only relative to observers and users.

Absolutely essential, then, to understanding the nature of the natural sciences is the distinction between those features of reality that are intrinsic and those that are observer-relative. Gravitational attraction is intrinsic. Being a $5 bill is observer-relative. Now, the really deep objection to computational theories of the mind can be stated quite clearly. Computation does not name an intrinsic feature of reality but is observer-relative, and this is because computation is defined in terms of symbol manipulation, but the notion of a symbol is not a notion of physics or chemistry. Something is a symbol only if it is used, treated, or regarded as a symbol. The Chinese room argument showed that semantics is not intrinsic to syntax. But what this argument shows is that syntax is not intrinsic to physics. There are no purely physical properties that zeros and ones or symbols in general have that determine that they are symbols. Something is a symbol only relative to some observer, user, or agent who assigns a symbolic interpretation to it. So the question, "Is consciousness a computer program?" lacks a clear sense. If it asks, "Can you assign a computational interpretation to those brain processes that are characteristic of consciousness?" the answer is: One can assign a computational interpretation to anything. But if the question asks, "Is consciousness intrinsically computational?" the answer is: Nothing is intrinsically computational. Computation exists only relative to some agent or observer who imposes a computational interpretation on some phenomenon. This is an obvious point. I should have seen it 10 years ago, but I did not.

REFERENCE

Searle, J. R. (1980). Minds, brains, and programs. *Behavioral and Brain Sciences, 3,* 417–457.

II THE PROPERTIES OF CONSCIOUSNESS

Antti Revonsuo
University of Turku, Finland

THE FRAGILE SMELL OF ROSE, THE PECULIAR ABOUTNESS OF THOUGHT, AND THE ELUSIVE SUBJECT OF CONSCIOUS EXPERIENCE

Imagine that you are admiring a sunset from the top of a hill. What kind of properties does such a typical conscious experience involve? Let us begin with the simplest, most atomic sensory components of your experience. Your color vision allows you to enjoy an incredible pallet of different tints and shades: The sky and the clouds near the western horizon burn in purple as the red-orange sun slowly sinks behind the skyline. If you look further up, you can see how the shades of the dusk sky change from purple to violet, to light blue, and eventually to very dark, velvet blue, revealing the first diamond-like stars against the dim background. Maybe you can hear the sound of the waves and the screech of seagulls from the distance and smell the salty air rising from the ocean. The rock on which you are sitting feels hard and still slightly warm, but the chilling wind sends cold shivers down your spine and, eventually, you feel how you are getting goose pimples all over your skin. It is exactly these kinds of basic sensory components of experiences—defining the elemental ways things appear to a subject—that philosophers call *qualia*.

Of course, your experience of the night falling and the stars coming out is much more than a kaleidoscope of chaotic sensations. Most of your perceptions are carefully organized, thus forming contentful wholes. It is almost impossible for you to see the red-orange circle in your visual field as an uninterpreted or meaningless patch of color—you instantaneously categorize it as "the sun." That is, your perceptual and other mental states have *contents* that can be specified only by referring to things outside of yourself: the sky, stars, ocean, trees, and so on. Your conscious experience or thought is, in some way, about these external things, without itself being identical with them. This "aboutness" of mental states is what philosophers call their *intentionality.*

Furthermore, you yourself are related to your conscious experiences in a way in which no other person, creature, or thing in the whole universe is related to them. You can try to verbally describe the features of your experience to somebody else or you can show a picture of the place where you were when enjoying the experience, hoping that it will be able to produce, for your friends, an experience at least slightly reminiscent of your own. But even if you should share the very event itself, there and then, with your closest companion, you could not literally share your experience. Although you can bring your verbalizations or photographs into public, you cannot do the same for the conscious experience itself. It is not a phenomenon to be measured, observed, or experienced in public. Only you yourself know what the experience was like from your particular point of view. Maybe sunsets activate childhood memories of countryside summers for you or maybe romantic memories of your first kiss. Or maybe you tend to conceptualize a sunset as "the turning of the third planet of a small yellow sun in a distant arm of an average spiral galaxy so that the edge of the planet's own shadow is sweeping over the place where I am standing on the planet's surface."

That there is something that it is like to be you having that experience—something that the experience seems like from your unique vantage point, something intrinsic that is concealed from the outsiders, philosophers call the *subjective character of experience.*

These three properties of conscious experiences—qualia, intentionality, and subjectivity—have been a pain in the neck for philosophers of mind for longer than they would care to admit. The world as conceptualized in physics, chemistry, or biology does not seem to contain any properties comparable to these properties of the mental realm. From the standpoint of physics, the wavelengths of electromagnetic radiation we call "visible" are in no respect special—the radiation itself is similar to that on any arbitrary band of wavelengths. Qualia cannot be found in physics, and the same seems to be true of intentionality. No physical particle or object attempts to be *about* some other thing—physical objects simply do not refer to other things outside of themselves. Water does not refer to lakes, seas, ice, or drinking. Sticks do not refer to trees, nor are stones about rocks. Physical things are quite content as they are. Neither is there any hidden vantage point for a stick or a stone. There is no subjective character of being a stone, no world-as-experienced from the point of view of the stone.

It inevitably seems that these properties are indeed properties of the mind. But if we believe that the mind is, after all, the brain, and the brain is a physical object in the world, we cannot escape the conclusion that at least some physical objects in the world have to be able to house properties such as qualia, intentionality, and subjectivity. But how dare we postulate such extraordinary characteristics to the brain without at the same time having to admit that we have no idea how what we postulate could be possible? And if we have to admit that much, is it not almost the same as admitting that the brain as a distinctively physical object could not conceivably perform the trick, after all. Soon we may find that we have painted ourselves into a corner, the only escape from which is an uninviting path leading to dualism of some sort.

The really bad news comes from those philosophers who claim that our present failure to scientifically understand the subjectivity of consciousness is not due to just bad luck or lack of empirical data. Rather, they argue, subjectivity is irreconcilable with a scientific approach to phenomena in general. There undoubtedly is a wide variety of thoroughly alien minds somewhere in the universe—from bat and dolphin minds to inconceivable extraterrestrial beings. Their subjective, experiential worlds are presumably completely beyond our reach. But, we may wonder, should not science one day be able to tell us which kind of a subjective consciousness an alien mind enjoys? Or, vice versa, could an alien mind ever find out what it is like to be a human being? Scientific explanation of ordinary physical phenomena leads to a more general, objective, and precise account of the real nature of things. Thus, lightning or rainbows can be characterized by referring to concepts and properties that are not dependent on a specifically human (or any other creature's) standpoint: The concepts of science are accessible also to other than human minds. The point is, then, that we move from how objects appear to us to what those objects really are behind appearances. But how could this scheme be applied in the case of subjective consciousness? How could we move away from the way conscious mental states appear to us to something else that supposedly is the real nature of those states? Thomas Nagel (1974) argued that such a fatal move would not in the least take us any closer to the understanding of those states. Distancing away from the way conscious states appear to the subject would not help us to comprehend what it is like to have those states. Quite on the contrary, the subjectivity of the states vanishes like a mirage when we approach it armed with our conventional scientific weapons. Science simply does not have the ghost of a chance of capturing subjective consciousness and its elusive properties alive.

Maybe, however, the difficulty of reconciling these problematic properties with the scientific view of the world is a symptom of us depicting the problem in an inappropriate way. Maybe we have been lured by our commonsensical intuitions, once again, to believe in things that do not exist anywhere outside our own imagination. Are there really qualia? Perhaps instead of them there are some much less mysterious properties that we mistakenly take for qualia. And maybe aboutness is not a property confined to the mind, after all—maybe it is just a matter of degree whether one thing stands for another. And, who knows, perhaps even the idea that conscious beings inhabit a subjective, phenomenological world-as-experienced is a mistaken intuition. If we cannot give any genuine explanations, why not simply explain the annoying properties away. This certainly is the path chosen by many philosophers of mind. Nevertheless, it seems that the way a philosopher treats qualia, intentionality, and subjectivity is strongly interconnected with the way the same philosopher theorizes about consciousness in general. Therefore, we have to grasp the underlying philosophical paradigm to understand the solutions offered to individual problems such as problems concerning the properties of consciousness. And we have to inspect whether the offered philosophical theory of consciousness is compatible with empirical findings and empirically based models of consciousness.

The cornerstone of Daniel Dennett's approach to these properties of consciousness is to construct a theory from the third-person point of view, because "all science is constructed from that perspective" (Dennett, 1991, p. 71). Consequently, he denies that conscious experiences could have any subjective qualitative properties such as the way a wild strawberry tastes to person P1 at time t1 or the way the colors of the sky appear to person P2 at time t2, especially if it is implied that these properties can be completely isolated from the person's dispositions to believe or behave (Dennett, 1988, p. 45). That is, all the relevant discriminative states can be, according to Dennett, thoroughly characterized from the third-person point of view. Any systems, man or machine, exhibiting identical discriminative powers also share all conceivable "mental" properties: There are no qualia overhanging the dispositions to believe and behave. Thus, a wine-tasting machine that is capable of making the same discriminations as a human wine-taster does not miss any qualia.

This analysis of qualia means that Dennett also denies the possibility of a zombie that would be behaviorally indistinguishable from a normal human being and yet a nonconscious automaton. It is impossible to abolish any mental or experiential properties of a system while at the same time preserving the discriminative powers and behavioral dispositions. Furthermore, the idea of even adjusting the qualia while keeping the dispositions to react constant is, for Dennett, self-contradictory. In other words, he also denies the possibility of behaviorally undetectable "qualia-inversions."

However, such a denial of qualia is by no means universally accepted. There are many who believe that conscious experience does have qualitative properties that are not captured by any kind of description made from the third-person point of view: The subjective, first-person characteristics of consciousness simply are different from the third-person features. There is something "ontologically subjective" about consciousness (Searle, 1992), and we just have to learn to live with that fact. This does not necessarily mean that we can have no empirical means of getting at the qualitative properties of experiences. If there is a difference at the level of phenomenology (say, we experience colors differently although our behavior is identical), then there must be a difference at the level of the causal basis of this phenomenology, namely at the level of neurophysiology. If there are no relevant differences at that level, we may safely assume that similar causes have similar effects and that organisms constructed the way that we are *must* enjoy a phenomenology similar to ours. Thus, although the qualia themselves may be subjective, it does not mean that we cannot empirically discover facts concerning them.

Another comparable dispute concerns intentionality and its relation to consciousness, and this disagreement too divides philosophers of mind into fervently opposing camps. John Searle has repeatedly argued that there is a profound distinction between intentionality that is genuinely mental and other forms of intentionality. States and processes that are in principle not thinkable or experienceable (i.e., which are not potential contents of consciousness) cannot have intentionality in its genuinely mental form, which Searle calls *intrinsic intentionality.* When you have a perception, a belief, a mental image, or thought about a sunset, you have a state or process somewhere inside your brain that is thinkable or experienceable—in other words, it has intrinsic intentionality. But if you write a story or draw a picture of the sunset on paper, or if you program a simulation of it in your personal computer, are these things (pieces of paper with ink marks or running computer simulations) in themselves *about* the sunset in the same sense as your mental states and processes are? No. The text, the picture, and the program as physical states or processes, if left on their own, have nothing to do with sunsets. They are not about anything, they just are what they are, like any ordinary, well-behaved sticks and stones around us. However, a conscious being with intrinsically intentional mental states has the wonderful ability to *interpret* the ink marks on the paper or the show on the computer screen as being about a sunset. Thus, physical objects devoid of intrinsic intentionality in themselves can facilitate bringing about mental states and processes with the content "sunset" in the conscious, external interpreter. But still, the intentionality of the pictures and computers lies entirely in the eye of the beholder who is able to treat the systems devoid of intrinsic intentionality in a way that brings about experienceable mental states in the observer. The intentionality of the external objects is only in the world-as-experienced-by-the-subject, not in the world-as-described-by-physics. In a world where there are no conscious beings (and consequently, no experienceable states or processes), everything is what it is and nothing is about anything external to itself.

Not surprisingly, many if not most philosophers of mind do not accept such a radical dichotomy of intentionality into "the real thing" and other, lesser forms that exist only as derivatives from the intrinsic, genuine, and original form of intentionality. For example, Dennett strictly denies that mental representation or intentionality would be somehow different from all other representation, and he explicitly rejects the idea of intrinsic intentionality: "There is no such thing as intrinsic intentionality, especially if this is viewed, as Searle can now be seen to require, as a property to which the subject has conscious, privileged access" (Dennett, 1987, pp. 336–337). "A shopping list in the head has no more intrinsic intentionality than a shopping list on a piece of paper" (Dennett, 1990a, p. 62). According to Dennett, there is only derived intentionality. That is, all aboutness, all contentful states, are interpretations made from the third-person's "intentional stance," no matter whether the system to be interpreted is a text, an artifact, or a person (Dennett, 1990b). And in each case, there is no deeper form of intentionality to settle the question, "What is the real content of this system's intentional states, as opposed to what we hypothesize from the outside them to be?"

It seems that if we accept the existence of intrinsic intentionality, the whole problem boils down to consciousness. If we do not accept it, even entities that intuitively have nothing comparable to our mental contents must be allowed to have intentionality in exactly the same sense as we do in our conscious experiences.

In chapter 5, John Haugeland criticizes both Dennett's and Searle's theories of intentionality. He proposes that there is, in fact, much more common ground for these two philosophers' views than one is inclined to think at first glance. The key to this insight is to note the consensus between these philosophers that intentionality is normative. Haugeland argues that underlying this normativity is what he calls "commitment to constitutive standards" or "adopting a stance" (chapter 5, this volume, p. 121). However, Haugeland's account of adopting a stance is radically different from Dennett's. In Dennett's view, an external observer adopts the intentional stance towards an intentional system. Haugeland, by contrast, thinks that the stance must be adopted by the system itself: This is a prerequisite to the intentionality of the system. From this innovative characterization of the essence of intentionality, Haugeland proceeds to distinguish a new, intermediate variety of it—*ersatz* intentionality. This notion applies, for example, to animals and computers, which do not have any intrinsic intentionality at all. Thus, Haugeland's account parts company with Dennett's and Searle's at least in this important sense: Among biological and artificial systems (at least among those we know about), humans stand out as unique, possessing genuine intentionality. Robots and animals lack this property; they only have ersatz intentionality. In contrast to this, Dennett thinks that all these systems have intentionality in exactly the same sense, whereas Searle blesses humans and animals with genuine, intrinsic intentionality but discards robots and computers (if their intentionality is supposed to be due to the programs they implement).

In chapter 6, Daniel Dennett sets out to defend his view that there are no intrinsic subjective qualia over and above the discriminative states of a system and its dispositions to behave. He invites us to consider what would go on inside a color-detecting robot if it were given the task to compare, in its memory, different colors of things (e.g., the pinkness of the flamingo with the pinkness of the Pink Panther). Now, is it possible to put the robot in a state of "spectrum-inversion"—the condition described in many classical arguments for the reality of qualia? Dennett argues that the intuition according to which qualia-inversion is possible depends on an ungrounded (false) presupposistion that there is a single, central register or place that determines the way color x seems to person (or robot) y at time t. And, as we know from Dennett's multiple drafts model of consciousness, he flatly denies the possibility of such a center of consciousness in the brain.

Arguments concentrating on the properties of consciousness have made philosophy of mind somewhat of a branch of science fiction literature. The defenders and opponents of qualia, intrinsic intentionality, and subjectivity have allied themselves with all kinds of alien beings: extraterrestrials, robots, computers, zombies, and brains-in-the-vat. The purpose of such fantasies is to test our intuitions in unfamiliar circumstances in order to extract the essence of the properties of consciousness out into the open. Consequently, the way a philosopher treats the possibility of, say, zombies, often tells us how the same philosopher treats consciousness and its properties.

A zombie is a nonconscious system that lacks qualia, intrinsic intentionality, and subjectivity, but that nevertheless is able to exhibit succesful humanlike behavior or otherwise very sophisticated and flexible organism-environment interaction. A zombie cannot in any way sense or experience that it exists. Because Dennett does not accept qualia, intrinsic intentionality, or first-person phenomenology as something logically independent of discriminative powers and behavioral dispositions, it is obvious why he cannot grant the possibility of zombies, either. He thinks that the concept of a zombie is just incoherent (Dennett, 1982), which it is, of course, if the existence of consciousness is defined in terms of observable behavior. In consequence of such a definition, zombie is a being that both has consciousness (by its behavioral manifestations) and does not have consciousness (by the definition of a zombie).

Nevertheless, zombies have done a lot of valuable work for those who have defended the special status of consciousness and its properties against philosophers who claim that one or another branch of materialistic philosophy of mind has already been able to exhaustively explain mentality. Ned Block's (1980) famous "absent-qualia" arguments against functionalism concentrated on showing that a homunculi-headed system constructed to simulate the functional organization of a person and his behavior is a mere unconscious zombie. It is questionable whether such a system has any mental states at all, not to mention qualia. Probably the most celebrated philosophical argument in the history of cognitive science, John Searle's (1980) Chinese room argument, aimed at showing that a computer simulation of human understanding does not house any genuine intentional states. Any system defined purely syntactically, like computer programs, cannot have contentful mental states in virtue of the formal program they implement. In other words, computer simulations of human cognition may produce complex input–output relations and behavior, but not consciousness or intrinsic intentionality. At best, we end up with an army of unconscious zombies, once again.

Those who believe that zombies are indeed possible point out that behavior and functional roles are simply irrelevant to the existence of consciousness and its properties. Consciousness is a phenomenon completely independent of any conceivable behavioral outputs. Consequently, it is (logically) possible to have a full behavioral repertory coupled with zero consciousness (zombies) and full consciousness with zero sensory inputs and zero behavioral outputs (dreams, especially lucid dreams). However, there still remains a variety of zombies that even friends of qualia and intrinsic intentionality would deny. Consider two beings that are not only behaviorally but also *neurophysiologically* identical. The claim that only one of them is genuinely conscious and the other an unfortunate zombie is presumably not accepted even by the fondest supporter of zombies. Consciousness, it appears, could conceivably be separated from behavior and from discriminative powers, but there is no room to drive a wedge between consciousness and its causal basis, the brain. If there are no differences between two systems at the level of the brain, how could there be differences at the level of consciousness? If we accept the general principle that the mental supervenes on the physical, we cannot coherently claim that an anatomically indistinguishable duplicate of a normal human being would not be conscious. Same causes entail same effects, and identical microlevel organization entails identical macrolevel organization.

If you deny the logical possibility of all kinds of zombies (behaviorally and/or neurophysiologically indistinguishable ones), your theory of consciousness is going to look quite different than if you deny only the possibility of a neurophysiologically indistinguishable zombie but admit the possibility of a behaviorally indistinguishable one. Dennett was inclined to be of the former opinion concerning zombies, whereas Searle, for example, would probably be of the latter. And, as we all know, the theories of consciousness put forward by these two philosophers stand worlds apart.

In conclusion, the problems of qualia, intrinsic intentionality, and subjectivity have a common undercurrent, consciousness. This fact went unnoticed in the philosophy of mind for a long time. But now that we have realized it, we may always reflect philosophical accounts of these problems as revealing an implicit philosophical theory of consciousness. Whenever some philosopher or some new branch of the philosophy of mind declares that these problems have been solved or shown to be artificial pseudoproblems stemming from our confused everyday intuitions, we can always put the proposed theory into a test that we may call the *Zombie Test*. It goes like this: Imagine a creature that, according to the novel theory, minimally satisfies the description of an intelligent or mental system. Then ask: Do we have good, independent reasons to believe that this system has qualia, intrinsic intentionality, or subjectivity (if you believe that conscious beings ought to have such things)? Or does the creature thus depicted rather look like one more potential member in the Club of Philosophical Zombies, whose membership is rapidly increasing anyway? If the answer to the latter question is in the affirmative, you can reassure yourself that despite the claims to the contrary made by the supporters of the theory, the occasion very likely was not the last time that you ever heard of the problematic properties of conscious experiences.

REFERENCES

Block, N. (1980). Troubles with functionalism. In N. Block (Ed.), *Readings in philosophy of psychology* (pp. 268–306). Cambridge, MA: Harvard University Press.

Dennett, D. C. (1982). How to study human consciousness empirically or nothing comes to mind. *Synthese, 53,* 159–180.

Dennett, D. C. (1987). *The intentional stance*. Cambridge, MA: MIT Press.

Dennett, D. C. (1988). Quining qualia. In A. J. Marcel & E. Bisiach (Eds.), *Consciousness in contemporary science* (pp. 42–77). New York: Oxford University Press.

Dennett, D. C. (1990a). The myth of original intentionality. In K. A. Mohyeldin Said, W. H. Newton-Smith, R. Viale, & K. V. Wilkes (Eds.), *Modelling the mind* (pp. 43–62). Oxford: Clarendon Press.

Dennett, D. C. (1990b). The interpretation of texts, people and other artifacts. *Philosophy and Phenomenological Research, 50(Supplement),* 177–194.

Dennett, D. C. (1991). *Consciousness explained.* Boston: Little, Brown.

Nagel, T. (1974). What is it like to be a bat? *The Philosophical Review, 83,* 435–450.

Searle, J. R. (1980). Minds, brains, and programs. *Behavioral and Brain Sciences, 3,* 417–457.

Searle, J. R. (1992). *The rediscovery of the mind*. Cambridge, MA: MIT Press.

5 Understanding: Dennett and Searle

John Haugeland
University of Pittsburgh

RECONCILIATION

I want to attempt here what may seem the impossible: to agree with both Dan Dennett and John Searle about the mind—or, at any rate, about *intentionality.* I do not mean agreement on every detail, of course; that would be too much to hope. Rather, the idea is to outline a view that accommodates (what seem to me) the most central and the most important of their respective insights and intuitions, at the expense of a few others. As it happens, the effort to achieve unity entails the occasional point of disagreement with both of them as well as the introduction of some exogenous material that I can only hope is compatible with what's already there.

We can begin with some obvious and acknowledged common ground. Dennett and Searle agree in regarding intentionality as an entirely natural phenomenon and in their commitment to a scientific understanding of nature. Moreover, they are each materialists, at least to the extent of holding that (as a matter of fact, so far as we know) matter, suitably arranged and interacting, is necessary and sufficient for intentionality. In other words, two worlds that were materially identical would be mentally identical as well; but if you took away the matter or sufficiently rearranged it, you would destroy all intentionality too. On the other hand, neither Dennett nor Searle is sympathetic to traditional physicalist reductionism. That is, neither holds out any prospect for strict definitions of mental or intentional concepts in physical terms.

Where they diverge, then, is in the way they characterize that "suitable arrangement and interaction" of matter, and hence in their accounts of which systems might have it and how we can tell. The appearance of conflict, however, is artificially heightened by overemphasis on a few strategic examples—particularly computers and artificial intelligence (AI). Better, I think, to concentrate on basic principles as much as possible, with the hope that controversial cases will not simply be decided but illuminated. The ideal outcome would be not an adjudication but an explanation of why those cases are controversial in the first place (i.e., why they pull in both ways). But, before we can make any headway with that, we need a brief overview of the two approaches under consideration.

DENNETT ON INTENTIONALITY

Dennett introduced his notion of the *intentional stance* to articulate his view that intentionality is, in a certain sense, in the eye of the beholder. The sense is this: A system has intentional states—paradigmatically, beliefs and desires—just in case its behavior exhibits a specific sort of observable pattern (Dennett, 1987, pp. 13–35). The intentional states are not the same as that observable pattern or at least not the observable part of it; rather they are a kind of completion of the pattern that is more or less necessary for it to stand out clearly in the first place. So it is roughly as if you were given every other letter of a text or a scattered fraction of an image, and you "fill in" the "missing" pieces. Without that filling in, the visible part seems irregular and disjointed; yet its structure becomes conspicuous and compelling, once the remainder is interpolated. Furthermore, in terms of the new whole, the original part can itself be redescribed (i.e., in intentional terms).

It is important to understand, however, that this is not "inference to the best explanation." Dennett is *not* saying that the rest of the pattern is "really there" though, for some reason, invisible—so we must infer it from the part we can see. In particular, he is not saying that beliefs and desires are determinate brain configurations or processes which we are so far unable to observe. Rather, intentional states and processes are in principle nothing but our projected filling in of the pattern in such a way that it makes sense overall. This is the way in which they are "in the eye of the beholder"—or perhaps it would be better to say, in the sense of the understander. But note well: This by no means renders them fictional, gratuitous, or arbitrary. In any given case, the attribution of intentional states is strongly constrained, both by principles and by facts—so strongly, indeed, that it is a nontrivial achievement to succeed at it at all (Dennett, 1987, pp. 24ff).

The essential principle of the intentional stance is a constitutive standard of rationality. This, in effect, defines the kind of sense that can be made of a behavior pattern by interpolating intentional states and processes into it and simultaneously redescribing the behavior itself in intentional terms (e.g., as perceptions and actions). Rationality is a global property of a system's behavior in context, including actual truth and success (Dennett, 1987, pp. 17ff). Moreover, it must be understood as projectable (in Nelson Goodman's sense, that is, it supports counterfactuals). That is why it is a constraint on a pattern and a strong enough one to support interpolation and a new descriptive vocabulary. Perception and action make sense as, and only as, rational. The intentional stance is the descriptive and projective stance adopted by someone—the observer—who by counting on this pattern and using its vocabulary, understands the system as intentional (rational).

SEARLE ON INTENTIONALITY

Searle will have next to none of this. Above all, he rejects the idea that intentionality is in any sense in the eye of the beholder (Searle, 1992, pp. 7, 78). It is just as real—right there in the brain—as any other higher-order property of complex systems such as liquidity, digestion, photosynthesis, or life itself (pp. 14, 28ff). The microstructure of the brain is causally responsible for its mental (and any other emergent) properties in just the sense that the microstructure of water is causally responsible for its liquidity—though, of course, in a much more complicated way. This offers a perfectly straightforward solution to the traditional problem of mind–body interaction, while preserving certain other attractive features of that tradition. Thus, because mental states and events just are physical states and events, only described at a higher level, there is no mystery about how they can interact causally with other physical states and events. At the same time, however, these interactions are entirely contingent and are in no way constitutive of the mental as such.

So, for instance, it would be metaphysically possible—though appalling—to have mental states and events without any actual connection to perception or behavior. By the same token, it makes perfect sense—though it strains credulity—to imagine a system that behaved exactly as if it had mental states but in fact had none (Searle, 1992, pp. 65–71). This is the basis of Searle's rejection of the Turing test. It is not that he believes that AI is on the verge of producing systems that act just like people, even though they are zombies "inside." There are plenty of reasons to be skeptical of that. The point is rather that the Turing test—and by implication, the intentional stance—is incompetent to decide. Intentionality is, in Searle's terms, *intrinsic* to the systems that have it. No amount or kind of behavior is either necessary or sufficient.

Finally, Searle emphasized, in a way that Dennett does not (and cannot), that intrinsic intentionality is essentially *subjective* (Searle, 1992, pp. 93–100). Obviously, subjective does not here mean observer-relative or in the eye of the beholder as when we say that aesthetic tastes are subjective. Rather, it means that intentional states—again, paradigmatically beliefs and desires—always belong to somebody, a subject who has them. Subjectivity, according to Searle, entails consciousness in the following way: Any genuine subject, at least sometimes, has some of its intentional states consciously and could, in principle, have any of them consciously. That is, no system incapable of consciousness is a subject or has any intentional states (except derivatively).

Thus, in Searle's terminology, the difference between him and Dennett is that he regards intentionality as intrinsic and subjective, whereas Dennett regards it as observer-relative and objective. Subject to some qualifications about how these terms are to be understood, I think this characterization is fair enough; and, moreover, put this way, I think Searle is closer to the truth. Predictably, however, the qualifications I have in mind are not altogether trivial nor necessarily acceptable to Searle. Indeed, as my introductory remarks hinted, the net effect is to make Searle's position rather closer to Dennett's than either of them might expect (or perhaps appreciate).

NORMATIVITY AND ASPECT

Everyone agrees that intentionality is somehow normative. For Dennett, this comes out in the appeal to rationality as the criterion for adequate interpretation (Dennett, 1987, pp. 51–54, 342–344). The point, of course, is not that the interpreter must be rational but rather that the system being interpreted must, as interpreted, be rational. That is, the normative standard is imposed on the intentional system as such. Still, it is being imposed—that is, it is observer-relative. For Searle, the normative element shows up in what he called the *satisfaction conditions* for intentional states, and needless to say, he maintained that the having of satisfaction conditions is itself also intrinsic to intentional mental states (pp. 51, 238). But, for both authors, the mental just *is*—or, at any rate, supervenes on—the physical; and Searle has no more to say than Dennett about how a physical system might have normative properties intrinsically. In fact, Dennett might well claim it as an advantage of his "eye-of-the-beholder" approach that it need not address this question (or, anyway, so it seems). Searle, on the other hand, cannot get off so easily.

To say that intentional states are normative (i.e., are subject to norms) is to say that there is a way that they are supposed to be that may or may not be the way they are. Moreover, it is characteristic of the specific normativity of the intentional that whether any given state is as it is supposed to be depends on some determinate condition or state of affairs outside of it—namely, its satisfaction condition (sometimes called its *intentional object*). Finally, however, all intentional states (that we know of) are finite; they do not and cannot depend on everything about any concrete particular but can only depend on specific aspects of it. Searle called this the *aspectual shape* of intentional states.

For example, I believe that Mugsy is a brown dog. As an intentional state, my belief is subject to norms; more specifically, as a belief, it is supposed to be true. Whether it is in fact true, however, depends on something outside of it—namely, Mugsy. But it does not and cannot depend on everything about Mugsy, because Mugsy has infinitely (or, at least, indefinitely) many concrete determinations, and I am incapable of beliefs with infinite determinate content. Instead, my belief depends on only two of Mugsy's determinations: her color and her biological species. That is, it depends only on whether she is brown and a dog; this is its *aspectual shape*. (The term aspectual shape suggests a visible "look"; and Searle did associate it with "seeing as," "perspective," and "point of view." But I think the argument from finitude that Searle does not give is both more general and more fundamental.)

A SPECIAL CASE: CHESS PERCEPTION

So the question is: How can naturally evolved physical brain configurations be normative in this way intrinsically. This is the question that I want to sketch an answer to—an answer that superficially sides with Searle against Dennett but more deeply brings the two of them together. I begin by considering a special case and a special problem concerning it—a problem that I try to solve in the special case. Then I suggest that the case, the problem, and the solution all generalize. The special case is visible chess phenomena: pieces, positions, moves, and so forth. The special problem is how these phenomena can be the satisfaction conditions of perceptual states.

Why is that a problem? Well, suppose that I am looking at the current position, and I see that your knight is threatening my rook. In virtue of what is it the case that the content of my perception is what it is—that is, has those satisfaction conditions? Science, of course, has learned a lot about vision. We know, for example, that ambient photons bounce off the surfaces of the knight and rook, enter my eyeball, and stimulate a complex pattern of pulses in my optic nerve that ultimately causes my perceptual state. Moreover, had my bishop been where my rook is, then the photon and pulse patterns would have been different, and most likely I would have seen your knight as threatening my bishop instead.

So causal histories are highly relevant to what we perceive. Nevertheless, they cannot be what determine the content of our perceptual states, for two closely related reasons. In the first place, the pieces on the board are only one stage in those causal histories, and nothing in the histories themselves picks out that particular stage as more important than the others, even counterfactually. That is, the causal histories as such give no reason to say that I perceive the chess pieces, as opposed to the photons or even the pulses in my optic nerve. Second, and more to the heart of the matter, causal histories are not normative. The content of my perception, its satisfaction condition, is what is supposed to have caused it, regardless of what actually caused it. Misperceptions and illusions, after all, are caused too—just not by what they purport to be of. (Deviant causal chains complicate things even more; but they can be set aside in the present context.)

What we need, then, is an account of why my perception is supposed to have been caused by your knight threatening my rook, and why that normative condition is intrinsic—that is, not observer-relative. (Such an account would also explain why that stage in the causal history is the relevant one.) To understand the normativity of perception, we must understand what perception is and what is required of it. This is where the artificiality of the chess example helpfully clarifies and simplifies matters. Perceiving chess phenomena is clearly an integral part of playing chess—along with making moves, weighing options, developing strategies, and so on. (Of course, one can also be a spectator at a chess match; but that capacity, surely, is parasitic on the ability to play.)

CONSTITUTIVE COMMITMENT

Consequently, we must make a brief digression about what it is to play chess. Two points are important for what follows. First, chess phenomena cannot be identified with any particular shapes, colors, materials, or other low-level properties; rather, they are ontologically constituted by higher-order standards that are spelled out in the rules of the game. So, on the one hand, chess pieces do not have to be the familiar wooden or plastic figurines: They can be marks on a piece of paper, costumed servants, patterns of light on a computer screen, or what have you—just so long as they are moved consistently according to the rules of chess. And, on the other hand, the pieces of an ordinary chess set could easily be used to play some quite different game—in which case they would not be chess pieces or phenomena. Hence, whether there are chess phenomena there to be perceived at all depends on whether the standards constitutive for chess are being met.

Second, the chess player/perceiver cannot be indifferent as to whether these standards are being met. If I am to see something as, say, a knight threatening a rook, I must be committed to it being part of a larger pattern of other things I can or could see—other pieces, moves positions, and so forth—that consistently accord with the rules. This commitment, I believe, is precisely what Dennett calls a *stance*—not an intentional stance, of course, but rather a chess stance. The constitutive role of the rules of the game exactly parallels the constitutive role he assigned to rationality-in-context for the intentional stance. Without adopting such a stance (that is, committing to it), it would be impossible to see anything as a chess phenomenon at all.

Commitment to constitutive standards means at least three things. First, it means that if something does not seem to accord with them, then one had better double check to make sure there is no mistake. Second, if the anomaly persists and no mistake can be found, then one must either modify the standards or give up on them (and the corresponding perceptions) altogether. But, third, modifying or giving up on established constitutive standards is not a matter to be taken lightly. For instance, I might believe that you had made an illegal move if I mistook your bishop for a pawn or got confused about the icons on our new computer. But if something does seem illegal to me, I cannot just ignore it. I need to find (and correct) my mistake (if any), because if there is no mistake—if you are in fact making (what would be) illegal chess moves—then this is not chess. Either you are playing a different game or none at all. Finally, however, I should not give up too easily, because constituting a domain of phenomena according to standards is a nontrivial and valuable achievement—and, in the case of established standards, one with a history of reliability.

OBJECTIVITY (TRANSCENDENTAL PHILOSOPHY)

The stakes, however, are higher than just rule-governed games. The idea of constituted phenomena goes back to Kant. The idea that there can be multiple phenomenal domains constituted according to different and perhaps contingent standards is a staple of 20th century philosophy, as the litany of terms like *regional ontology, framework of entities, conceptual scheme, scientific paradigm*, and now stance should make obvious. Although these are not all equivalent, what they have in common is a conviction that the objects of perception, thought, and action are intelligible as the objects they are only in terms of some prior commitment on our part to the limits of what they can be. And the terms provided (constituted) by that commitment are precisely the aspects under which those objects can be the objects of intentional states.

And that brings us back to Searle and the normativity of perception. Suppose there is something odd about the pattern of photons bouncing off your knight into my eye. For some reason, from my current angle of view, that pattern is much more like what would usually come from a bishop than what would usually come from a knight. So I have the response that I would usually have if there were a bishop there. Is that a mistake? The question can be put more sharply. Have I misperceived your knight, or have I correctly perceived the photons? The account of constitutive standards provides the answer. What I am committed to is playing chess. Chess standards constrain pieces, positions, and moves on the chessboard—not photons. Accordingly, the questions of misperception and correctness that can arise (and must be resolved) in the context of chess concern only those pieces, positions, and moves—not the photons. This is why my perceptions are supposed to be of the pieces, positions, and moves—not the photons. In other words, *objectivity* as such is constituted.

SUBJECTIVITY (TRANSCENDENTAL PHILOSOPHY)

But is it intrinsic? Yes! Commitment to standards is the very foundation and essence of intrinsic intentionality. To see this, we must examine what is meant by intrinsic. Obviously, the intentionality of some particular belief or desire cannot be intrinsic to that individual state all by itself. That would be incompatible with the holism of intentionality and its essential dependence on the background—two theses that Searle (among others) has espoused for years (Searle, 1992, pp. 175–178). Rather, the intrinsicness must pertain at once to all the intentional states of a single system, and it means that the intentionality of those states is independent of any other intentionality, that is, the intentionality of any other system. For instance, printed marks on a page or the states of an adding machine may have various meanings but only because we assign those meanings to them. Their intentionality is derivative from ours, and in that sense observer (or user) relative. Intrinsic intentionality is not in that way derivative (pp. 78–82, 211ff).

That, however, is merely a negative characterization; it says what intrinsic intentionality is not. A genuine positive characterization would have to show how intentionality that is not derivative is even possible—that is, how it is possible for any system to have intentional states on its own. (This amounts to showing how intentionality is possible at all, because derivative intentionality only makes sense if there is some nonderivative intentionality for it to be derivative from.) It is worth remarking in passing that what counts as a single system is so far a free parameter in the account; Searle often speaks as if the relevant systems are individual brains, but that is not built into the theory. What matters is that the unity of the system follows from the same positive account of how such a system could have intentional states on its own, that is, intrinsically.

But that is precisely what commitment to constitutive standards provides. The unity of the system is the unity of a single consistent commitment in terms of which a plurality of intentional states can be normatively beholden to their constituted satisfaction conditions. Moreover, it is the basis of the necessary subjectivity of intentional states. 'Subject' here cannot just mean grammatical subject: Even an adding machine is the subject of its states in that sense. Rather, 'subject' means something like author and owner—someone who is responsible for the states in question and to whom they matter. Surely that subject is none other than the one who is committed to the very standards that render these states intentionally normative in the first place. My commitment to getting my intentional states right is what makes them my own. In other words, subjectivity as such is constituted.

It might even be that commitment throws some light on the relevance of consciousness, for one of the characteristic features of consciousness is that though it is not always reflective, it is always potentially reflective. The self is, so to speak, always there in the background. But commitment to constitutive standards likewise lurks always in the background. When I am playing chess, I am not constantly double-checking my perceptions to ensure their accuracy; but I am always quietly at the ready. I am always, so to speak, "on guard," so that whenever there is an anomaly, then the issue can come to the fore.

UNDERSTANDING

I have, in effect, been urging Dennett's hallmark notion of a stance onto Searle, as the foundation of his own account of intentionality. But I have not actually been urging Dennett's own position. For Dennett understands intentionality in terms of a particular stance—the intentional stance—adopted toward a system; whereas I am suggesting that we understand intentionality in terms of a stance—any stance—adopted by a system. Dennett himself discusses what he calls *higher-order intentionality*, which would have to involve a system adopting the intentional stance, adopting the stance that something else is adopting the intentional stance, and so on (Dennett, 1987, pp. 243ff, 269ff). But this is not quite the same as what I'm talking about, nor does Dennett make anything like the use I am trying to make of it.

Here is why I think adopting a stance is prerequisite to intentionality. Any number of different states and arrangements of matter can be regarded as carrying information about things other than themselves. This can be a consequence of natural causal dependencies, as when clouds inform us of imminent rain; or it can be a matter of tacit or explicit convention, as when a badge informs me that someone is a police officer. More interestingly, complex systems can be built that record, transmit, combine, and otherwise usefully manipulate such information-bearing states in ways that systematically preserve their status as reliable bearers of information. For instance, an automated business accounting system can be built such that its manipulations preserve correct arithmetic relationships (according to the relevant interpretations of its states), thereby extracting many useful results. This is what we now call *computation*.

What is conspicuous about all such systems, whether automated or not, is that they themselves do not have a clue about the information that (from our point of view) they carry. Clouds do not understand anything about rain nor do badges about police. Likewise, a computer has no understanding whatever about business, accounting, or even arithmetic. Understanding a domain and its entities is understanding the principles according to which that domain and those entities are constituted; and such understanding can be nothing other than a commitment to those principles. Intentionality presupposes a committed stance because intentionality—meaning—presupposes understanding.

Searle once said of computers, including AI—what I call Good Old Fashioned AI, or GOFAI—systems, that they understand exactly nothing; they are not in that line of business (Searle, 1980, p. 288). I like the straightforwardness of this claim; furthermore, I agree with it completely. GOFAI systems (including, of course, computer chess-playing systems) are utterly incapable of any sort of commitment: Nothing is at stake for them. They are never in a position of having to figure what their own mistake could have been, on pain of having to modify the standards, or give up the game. Therefore, they understand nothing.

For what it is worth, I have no reason to believe that connectionist or any other approaches to AI are any better off than GOFAI in this regard. But I make no principled claim as to whether artificial systems of any sort will ever understand. Who knows?

ERSATZ INTENTIONALITY

Nevertheless, I do not agree with Searle about computers and other systems entirely. It seems to me that he makes too few distinctions, in a way that results in some misclassifications. He describes three ways in which we talk about intentionality: (a) *intrinsic* intentionality; (b) *derived* intentionality; and (c) *as-if* intentionality. The first is the real thing; it is what we and the higher (i.e., conscious) animals share. The second is intentionality, but only delegated or borrowed from some prior intrinsic intentionality; it is what computers and public inscriptions share. For instance, a printed sign that says "No smoking" really does mean "No smoking" but only because we give it that meaning; it can mean nothing all by itself. The third category is not really intentional at all but only a sort of metaphor, as when we say that water seeks its own level, or the saplings are trying to grow up to the sunlight.

Now I have no problem (for the moment) with the classifications of people, signs, and water. But animals and computers are more difficult and might better be handled by an intermediate classification that I propose to call *ersatz intentionality*. Consider a GOFAI robot of a sort not too far from current technology. It wanders around campus avoiding obstacles, keeping track of which paths are blocked (so as to stay up to date on the quickest routes), recharging its batteries periodically, and perhaps delivering packages or recording how long each parked car has been where it is. This system not only contains a large number of representations, but it also modifies them in direct response to its environment, relies on them to help it reach various goals, manipulates them internally in various combinations to figure out what to do and how to do it—all without human intervention.

It seems to me that the relation of this robot to its representations is importantly different from the relation that a piece of paper bears to a message written on it. There is a clear (even if limited) sense in which the robot is actually making and using those representations as representations, that is, as aids to dealing vicariously with other things that are not themselves present. If the representations were different, the robot would behave differently; if they were false, it would perform more poorly (or perhaps correct them). None of this, or anything like it, is true of the piece of paper. So it does not seem to me that the robot and the paper should be classified together. (Here I am agreeing with Dennett while disagreeing with Searle.)

On the other hand, not only are all of the robot's resources and goals preprogrammed by its designers, but also and more importantly, all of the standards for what counts as an object, an adequate representation, a goal, a success, and so on, are tacitly presupposed and hard-wired in that design. The point is not that the standards are given to the robot; rather, the robot does not have them at all, they remain entirely external to it. So, even though its actual intentionality is not merely delegated (like that of the words on a page), nevertheless the standards in virtue of which those states can be understood as intentional and normative are conferred from the outside. Hence, its states do not belong to it (as a subject) in the way that ours belong to us. This is why I want an intermediate classification between genuine and merely derived intentionality—namely, ersatz intentionality. (Neither Dennett nor Searle acknowledges such an intermediate status.)

ANIMALS

Once ersatz intentionality is distinguished and characterized, however, it becomes apparent that there are candidates for it besides GOFAI robots. In particular, I suggest that (as far as we know) the intentionality of animals is entirely ersatz (except for purely tropistic creatures, whose intentionality is at best "as-if"). That is, we can understand animals as having intentional states but only relative to standards that we establish for them. This makes animal intentionality exactly analogous to biological teleology. We say that the "purpose" of the heart is to pump blood, that it is "supposed to" work in a certain way, that functional descriptions are "normative," and so on. This is not mere as-if teleology (mere metaphor or *façon de parler*); nor is it any delegation of our own purposes and norms—like those for tools (by many accounts). But finally, of course, the heart does not have any purposes in the way that a person does, nor does it accede to any norms on its own responsibility. Extending my new terminology, I could say that biological systems have only ersatz teleology and normativity.

Searle will agree, I think, with this characterization of biological teleology (see Searle, 1980, pp. 237ff). Dennett will not only agree (it is the design stance, after all) but accept the proposed affinity with animal intentionality. But there the congeniality ends. For Searle maintains that animal intentionality is intrinsic, just like our own (and unlike robots), whereas Dennett wants the intentional stance to apply to us, animals, and robots all in the same way—leaving no distinction for ersatz to mark. By splitting the difference between them, I abandon both.

By my lights, animal intentionality is ersatz because (or to the extent that) animals do not commit to constitutive standards, hence do not submit themselves to norms, and do not understand anything. This is not to say that they do not perceive, remember, desire, learn, strive, and so on, but only to say that it is all ersatz—just as it would be for advanced robots. The fact that higher animals are more similar to us than are robots physiologically, ethologically, and phylogenetically is not necessarily significant when considering intentionality. After all, animals are also very different from us in many other ways such as morality, tradition, art appreciation, commerce, technology, and so on. Indeed, in a few respects, like the employment of structured representations and certain kinds of problem solving, machines are, on the face of it, more our kin than are monkeys.

The point is that none of these observations carries any probative weight unless and until there is some prior account of which kinds of considerations are relevant. And that prior account could itself only be based in a philosophical explication of what intentionality is and how it works. I have sketched one possible explication in terms of commitment to constitutive standards. According to this approach, however, all of the previous comparisons are equally irrelevant. With regard to what matters for intentionality, robots and animals are more similar to each other than either is to us. Only people have intrinsic intentionality.

CONCLUSION

I suggest that there is more in common to Dennett and Searle on intentionality than meets the eye—including their eyes. Pivotal to the envisioned rapprochement is Dennett's basic notion of a stance but used in a way that neither he nor Searle anticipates. If a stance is conceived as a constitutive commitment, then it can be seen as a transcendental ground of objectivity, subjectivity, and normativity—all of which are prerequisite to intentionality. A committed stance is the essence of understanding—hence my title.

REFERENCES

Dennett, D. C. (1987). *The intentional stance.* Cambridge, MA: MIT Press.

Searle, J. R. (1980). Minds, brains, programs. *The Behavioral and Brain Sciences, 3,* 417–424. Reprinted in J. Haugeland (Ed.), *Mind design*. Cambridge: MIT Press (1982). (Page citations are to the reprint.)

Searle, J. R. (1992). *The rediscovery of the mind.* Cambridge, MA: MIT Press.

6 Instead of Qualia[1]

Daniel C. Dennett
Tufts University

Philosophers have adopted various names for the things in the beholder (or properties of the beholder) that have been supposed to provide a safe home for the colors and the rest of the properties that have been banished from the external world by the triumphs of physics: *raw feels, sensa, phenomenal qualities, intrinsic properties of conscious experiences, the qualitative content of mental states,* and, of course, *qualia,* the term I use. There are subtle differences in how these terms have been defined, but I am going to ride roughshod over them. I deny that there are any such properties. But I agree wholeheartedly that there seem to be.

There seem to be qualia, because it really does seem as if science has shown us that the colors cannot be out there and hence must be in here. Moreover, it seems that what is in here cannot just be the judgments we make when things seem colored to us. Do not our internal discriminative states also have some special intrinsic properties, the subjective, private, ineffable properties that constitute the way things look to us (sound to us, smell to us, etc.)? Those additional properties are the qualia, and before looking at the arguments philosophers have devised in an attempt to prove that there are these additional properties, I try to remove the motivation for believing in these properties in the first place, by finding alternative explanations for the phenomena that seem to demand them. Then the systematic flaws in the attempted proofs will be readily visible.

[1] Portions of this chapter are drawn on Dennett (1991a, 1991b).

An excellent introductory book on the brain contains the following passage: "'Color' as such does not exist in the world; it exists only in the eye and brain of the beholder. Objects reflect many different wavelengths of light, but these light waves themselves have no color" (Ornstein & Thompson, 1984, p. 55).

This is a good stab at expressing the common wisdom, but notice that taken strictly and literally, it cannot be what the authors mean, and it cannot be true. Color, they say, does not exist "in the world" but only in the "eye and brain" of the beholder. But the eye and brain of the beholder are in the world, just as much parts of the physical world as the objects seen by the observer. And like those objects, the eye and brain are colorful. Eyes can be blue or brown or green, and even the brain is made not just of grey (and white) matter: in addition to the *substantia nigra* (the black stuff), there is the *locus ceruleus* (the blue place). But, of course, the colors that are "in the eye and brain of the beholder" in this sense are not what the authors are talking about. What makes anyone think there is color in any other sense?

The common wisdom is that modern science has removed the color from the physical world, replacing it with colorless electromagnetic radiation of various wavelengths, bouncing off surfaces that variably reflect and absorb that radiation—due to what Locke and Boyle called their secondary qualities, dispositional powers composed of their basic, or primary qualities. It may look as if the color is out there, but it is not. It is in here. It seems to follow that what is in here is both necessarily conscious (otherwise it is not all the way in) and necessarily qualitative (otherwise color would be utterly missing in the world). Locke's way of defining secondary qualities has become part of the standard layperson's interpretation of science, and it has its virtues, but it also gives hostages: the things produced in the mind. The secondary quality *red*, for instance, was for Locke the dispositional property or power of certain surfaces of physical objects, thanks to their microscopic textural features, to produce in us the idea of red whenever light was reflected off those surfaces into our eyes. The power in the external object is clear enough, it seems, but what kind of a thing is an idea of red? Is it, like a beautiful gown of blue, colored—in some sense? Or is it, like a beautiful discussion of purple, just about a color, without itself being colored at all? This opens up possibilities, but how could an idea be just about a color (e.g., the color red) if nothing anywhere *is* red—intrinsically?

This reasoning is confused. What science has actually shown us is just that the light-reflecting properties of objects—their secondary qualities—cause creatures to go into various discriminative states, underlying a host of innate dispositions and learned habits of varying complexity. And what are the properties is these internal states? Here we can indeed play Locke's card a second time: These discriminative states of observers' brains have various primary properties (their mechanistic properties due to their connections, the excitation states of their elements, and so forth), and in virtue of these primary properties, they too have various secondary, merely dispositional properties. In human creatures with language, for instance, these discriminative states often eventually dispose the creatures to express verbal judgments alluding to the color of various things. The semantics of these statements makes it clear what colors supposedly are: reflective properties of the surfaces of objects or of transparent volumes (the pink ice cube, the shaft of limelight). And that is just what colors are in fact—though saying just which reflective properties they are is tricky.

Do not our internal discriminative states also have some special intrinsic properties, the subjective, private, ineffable properties that constitute the way things look to us (sound to us, smell to us, and so forth)? No. The dispositional properties of those discriminative states already suffice to explain all the effects: the effects on both peripheral behavior (saying "Red!", stepping on the brake, and so forth) and internal behavior (judging "Red!", seeing something as red, reacting with uneasiness or displeasure if red things upset one). Any additional qualitative properties or qualia would thus have no positive role to play in any explanations, nor are they somehow vouchsafed to us directly in intuition. Qualitative properties that are intrinsically conscious are a myth, an artifact of misguided theorizing, not anything given pretheoretically.

We do have a need, as David Rosenthal (1991) showed, for properties of discriminative states that are in one sense independent of consciousness, and that can be for that very reason informatively cited in explanations of particular contents of our consciousness. These properties are partially, but not entirely, independent of consciousness. We may call such properties *lovely* properties as contrasted with *suspect* properties. Someone could be lovely who had never yet, as it happened, been observed by any observer of the sort who would find her lovely, but she could not—as a matter of logic—be a suspect until someone actually suspected her of something. Particular instances of lovely qualities (such as the quality of loveliness) can be said to exist as Lockean dispositions prior to the moment (if any) where they exercise their power over an observer, producing the defining effect therein. Thus, some unseen woman (self-raised on a desert island, I guess) could be genuinely lovely, having the dispositional power to affect normal observers of a certain class in a certain way, in spite of never having the opportunity to do so. But lovely qualities cannot be defined independently of the proclivities, susceptibilities, or dispositions of a class of observers. Actually, that is a bit too strong. Lovely qualities would not be defined—there would be no point in defining them, in contrast to all the other logically possible gerrymandered properties—independently of such a class of observers. So although it might be logically possible (in retrospect one might say) to gather color property instances together by something like brute force enumeration, the reasons for singling out such properties (for instance, in order to explain certain causal regularities in a set of curiously complicated objects) depend on the existence of the class of observers.

Are elephant seals[2] lovely? Not to us. It is hard to imagine an uglier creature. What makes an elephant seal lovely to another elephant seal is not what makes a woman lovely to another human being, and to call some as-yet-unobserved woman lovely who would mightily appeal to elephant seals would be to abuse both her and the term. It is only by reference to human tastes that are contingent and indeed idiosyncratic features of the world that the property of loveliness (to a human being) can be identified.

On the other hand, suspect qualities (such as the property of being a suspect) are understood to presuppose that any instance of the property has already had its defining effect on at least one observer. You may be eminently worthy of suspicion—you may even be obviously guilty—but you cannot be a suspect until someone actually suspects you. The tradition that Rosenthal denied would have it that sensory qualities are suspect properties—their *esse* is in every instance *percipi*. Just as an unsuspected suspect is no suspect at all, so an unfelt pain is supposedly no pain at all. But for the reasons Rosenthal adduced, this is exactly as unreasonable as the claim that an unseen object cannot be colored. He claimed, in effect, that sensory qualities should rather be considered lovely properties—like Lockean secondary qualities generally. Our intuition that the as-yet-unobserved emerald in the middle of the clump of ore is already green does not have to be denied, even though its being green is not a property it can be said to have intrinsically. This is easier to accept for some secondary qualities than for others. That the sulphurous fumes spewed forth by primordial volcanos were yellow seems somehow more objective than that they stank, but so long as what we mean by yellow is what we mean by yellow, the claims are parallel. Suppose some primordial earthquake cast up a cliff face exposing the stripes of hundreds of chemically different layers to the atmosphere. Were those stripes visible? We must ask to whom. Perhaps some of

[2] In Dennett, 1991a and 1991b, I misremembered the name of this amazing beast, whose antics I watched on the California coast in 1980—I called them sea elephants. No one has yet bothered to correct me on that, so I correct myself.

them would be visible to us and others not. Perhaps some of the invisible stripes would be visible to pigeons (with their tetrachromat color vision) or to creatures who saw in the infrared or ultraviolet part of the electromagnetic spectrum. For the same reason one cannot meaningfully ask whether the difference between emeralds and rubies is a visible difference without specifying the vision system in question.

The same moral should be drawn about the sensory qualities Rosenthal attributes to mental (or cerebral) states. Like Lockean secondary qualities in general, they are equivalence classes of complexes of primary qualities of those states and thus can exist independently of any observer, but because the equivalence classes of different complexes that compose the property are gathered by their characteristic effect on normal observers, it makes no sense to single them out as properties in the absence of the class of observers. There would not be colors at all if there were no observers with color vision, and there would be no pains at all if there were no subjects capable of conscious experience of pains, but that does not make either colors or pains into suspect properties.

Rosenthal (in a personal communication) asked whether this is not too strong. Why should the existence of pains require subjects capable of conscious experience of pains, as opposed simply to subjects capable of having nonconscious pains? Fair question, and his implied point is a good one—except for what amounts, in the end, to a lexical quandary that can be brought out by considering the parallel with color. There is nothing except the specific effects on normal human beings that demarcates the boundaries of the visible spectrum. Infrared and ultraviolet radiation does not count as subserving color vision (at least according to a sort of purist definitional taste) even in creatures who respond to it in the ways we respond to the humanly visible spectrum. "Yes, it is like color vision, but it is not color vision," someone might insist. "Color vision is vision whose proper objects are (only) red through violet." Now imagine that we confront a set of primary property complexes as candidates for the secondary property of pain and suppose it is a somewhat enlarged set (it includes infrapain and ultrapain, in effect), including outlying cases of which we human beings would never be conscious (but that have the sorts of effects on variant human beings that paradigmatic pains have on us, etc.). Would those be pains? There would certainly be a property that was the property picked out by that set, but would it be pain? (Not a terribly interesting question.)

I claim, then, that sensory qualities are nothing other than the dispositional properties of cerebral states to produce certain further effects in the very observers whose states they are. It is no objection to declare that it just seems obvious that our mental states really do have intrinsic properties over and above their dispositional properties. (If this were a good argument, it would be a good argument against the original distinction by Locke, Boyle and others, between primary and secondary qualities, for it certainly seems obvious that physical objects have their color properties intrinsically—just look at them!) It does indeed appear that we somehow enjoy, in our minds, some sort of direct and intimate access to intrinsic properties of our conscious states, but as Rosenthal observes, "We need not preserve the 'element of truth' in erroneous commonsense intuitions when we become convinced that these intuitions reflect how things appear, rather than how they really are" (Rosenthal, 1991, p. 27).

The prima facie case for the common conviction that qualia are needed over and above the various merely dispositional properties of our cerebral states can be dramatized—and then exploded—by an example. We can compare the colors of things in the world by putting them side by side and looking at them, to see which judgment we reach, but we can also compare the colors of things by just recalling or imagining them in our minds. Is the standard red of the stripes on the American flag the same red as or is it darker or lighter or brighter or more or less orange than the standard red of Santa Claus' suit (or a British pillar box or the Soviet red star)? (If no two of these standards are available in your memory, try a different pair, such as Visa-card-blue and sky blue, or billiard-table-felt-green and Granny-Smith-apple-green, or lemon-yellow and butter-yellow.) We are able to make such comparisons in our mind's eyes, and when we do, we somehow make something happen in us that retrieves information from memory and permits us to compare, in conscious experience, the colors of the standard objects as we remember them (as we take ourselves to remember them, in any case). Some of us are better at this than others, no doubt, and many of us are not very confident in the judgments we reach under such circumstances. That is why we take home paint samples or take fabric samples to the paint store, so that we can put side by side in the external world instances of the two colors we wish to compare.

When we do make these comparisons in our mind's eyes, what happens? It surely seems as if we confront in the most intimate way imaginable some intrinsic subjective color properties of the mental objects we compare, but before trying to figure out what happens in us, let us look at a deliberately oversimplified version of a parallel question: What would go on inside a color-detecting robot given the same sort of task?

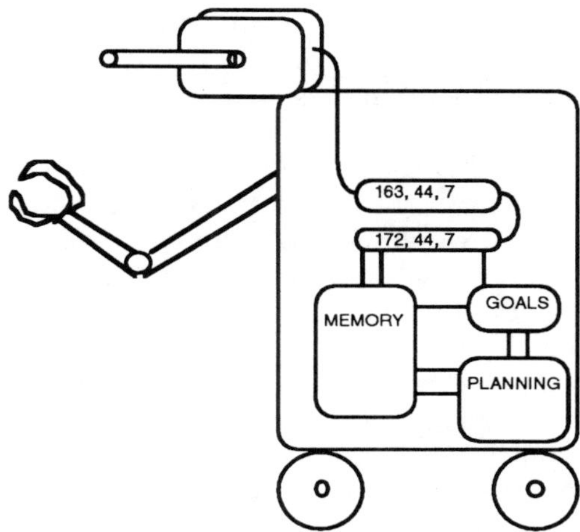

FIG. 6.1.

Figure 6.1 shows our robot, with a color TV camera feeding its signal into a sort of color CAD system that digitizes the images, color-coding each frame. No information is lost. Going to digital from analog simply makes for greater fidelity (as in the case of compact discs) and ease of explanation in the thought experiment. Suppose the computer can discriminate 1024 colors, numbered from 0 to 1023 (or in binary, from 0000000000 to 1111111111). It also codes, for every number, a saturation number between 0 and 127 and a brightness (or gray scale) between 0 (black) and 15 (white). Every value of these three variables is associated with a particular hue, intensity, and brightness, and to add a human touch, we can imagine the colors being assigned numbers not just in an orderly fashion but in the particular sort of order that human color vision would provide, with its red-green-blue (RGB) sensors in the retina, feeding into an opponent-process system (which gives us the complementary colors of afterimages). This is also easy to do, because there are off-the-shelf converters that recode the standard television scheme into the RGB system that computers use, in mimicry of the red-green-blue poles of maximal sensitivity in human vision. With our coding system in place, simple arithmetic operations preserve simple psychophysical effects: Plus-1 for brightness takes us to the nearest distinguishable shade of the same color, just changing its gray-scale value; minus-1 in hue changes the hue in the counterclockwise direction by one step, leaving saturation and brightness the same, and so forth.

Now suppose we place a color picture of Santa Claus in front of it and ask it whether the red in the picture is deeper than the red of the American flag (something it has already stored in its memory). This is what it would do: retrieve its representation of Old Glory from memory and locate the red stripes (they are labeled 163, 44, 7 in its diagram). It would then compare this red to the red of the Santa Claus suit in the picture in front of its camera that is transduced by its color graphics system as 172, 44, 7. It would compare the two reds by subtracting 163 from 172 and getting 9 that it would interpret as showing that Santa Claus red seems to it somewhat deeper—toward crimson—than American flag red.

This story is deliberately oversimple, to dramatize the assertion I wish to make: We have not yet been shown a difference of any theoretical importance between the robot's way of performing this task and our own. The robot does not perform its comparison by rendering colored representations and comparing their spectral properties, and neither do we. Nothing red, white, or blue happens in your brain when you conjure up an American flag, but no doubt something happens that has three physical variable clusters associated with it—one for red, one for white, and one for blue, and it is by some mechanical comparison of the values of those variables with stored values of the same variables in memory that you come to be furnished with an opinion about the relative shades of the seen and remembered colors. So although voltages in memory registers is surely not the way your brain represents colors, it will do as a stand-in for the unknown cerebral variables. In other words, so far we have been given no reason to deny that the discriminative states of the robot have content in just the same way, and for just the same reasons, as the discriminative brain states I have put in place of Locke's ideas.

But perhaps the example is misleading because of a lack of detail. Let's add some realistic touches. Each color in the robot's coded color space may be supposed to be cross-linked in the robot's memory systems with many earlier events in its experience—some of them positive, some negative, some directly, some indirectly. Let's make up a few possibilities. Because it is an IBM computer, let's suppose its designers have cleverly programmed in an innate preference for Big Blue. Any time Big Blue (477, 51, 5) comes in, it activates the "Hurrah for IBM!" modules that—if nothing else is going on—tend to get expressed in one way or another: Under extreme conditions, the robot begins to sing the praises of IBM, out loud. Under other conditions, the robot, if offered a choice, prefers to rest in a Big Blue room, exchanges an orange extension cord for a Big Blue one, and so forth. And, because there are plenty of other shades of blue, these near misses also tend to activate these IBM centers, though not as strongly.

Other colors will activate other routines (or subsystems, or modules, or memory clusters). In fact, every color gets assigned a host of disposition-provoking powers, most of which, of course, remain all but invisible in various slight reorganizations of its internal states. The effects may be supposed to be cumulative and associative or ancestral: A certain shade of rosy orange (950, 31, 3, let's say) reminds the robot of its task of turning on the porch light at sunset every night—a task long ago discontinued but still in memory. This activation of memory reminds the robot of the time it broke the light switch, shortcircuiting the whole house, and of the ensuing furor, the points subtracted from its good-designedness rating (a number that fluctuates up and down as the robot goes about its tasks and having the effect, when it is diminishing rapidly, of causing major chaotic reorganizations of its priority settings), and so forth.

In short, when a color is discriminated by the robot, the number does not just get stored away inertly in some register. It has a cascade of further effects, changing as a function of time and context.

Now along come the evil neurosurgeons—only these are computer scientists—and they do something ridiculously simple: They insert a little routine early in the system that subtracts each color's hue number from 1023 and sends on the balance (Fig. 6.2).

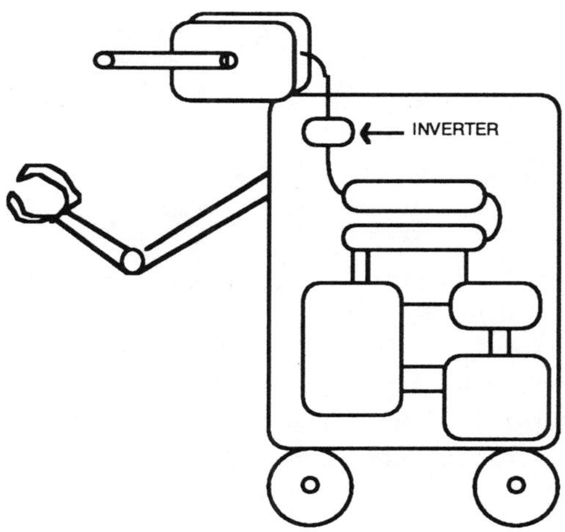

FIG. 6.2.

Spectrum-inversion! But to get the shocking effect, they have to make the change early before the numbers begin cascading their effects through the memory. If there is a central port, after the determination of color perception is accomplished and before all the effects, we have a place where we can locate the qualia of the system. An inversion there throws the system into bizarre behavior. It lunges for the bright yellow extension cord, and pale green light reminds it of that unhappy episode with the light switch.

But if some other computer scientists come along and also invert the numbers-for-colors throughout memory, the robot will behave in every way exactly the way it did before. It is a functional duplicate. But whereas Big Blue used to be registered by the hue number 477, now it is registered by the number 546, and so forth. Are the qualia still inverted? Well, the numbers are. And this would be detectable trivially by a direct examination of the contents of each register—robot B could be distinguished from robot A, its presurgical twin, by the different voltage patterns used for each number. But lacking the capacity for such direct inspection, the robots would be unable to detect any differences in their own color behavior or experience, and robot B would be oblivious to the switch from its earlier states. It would have to be oblivious to it, unless there were some leftover links to the old dispositional states that it could use for leverage—for identifying, to itself, the way things used to be.

But note that our spectrum inversion fantasy in the robot depends crucially on there being a central place from which all subjective color effects flow. It is the value in that register that determines which color something "seems to be" to the robot. But what if there is no such central depot or clearing house? What if the various effects on memory and behavior start to diverge right from the TV-camera's signal (as in Fig. 6.3), utilizing different coding systems for different purposes, so that no single variable's value counts as the subjective effect of the perceived color?

Then the whole idea of qualia inversion becomes undefinable for the system. And that is precisely the situation we now know to exist in human perception. For instance, in some cases of cerebral achromatopsia, patients can see color boundaries in the absence of luminance or brightness boundaries but cannot identify the colors that make the boundaries! (It is not that they claim to see in black and white either; they confidently name the colors and have no complaint about their color vision at all, but their naming is at chance!)

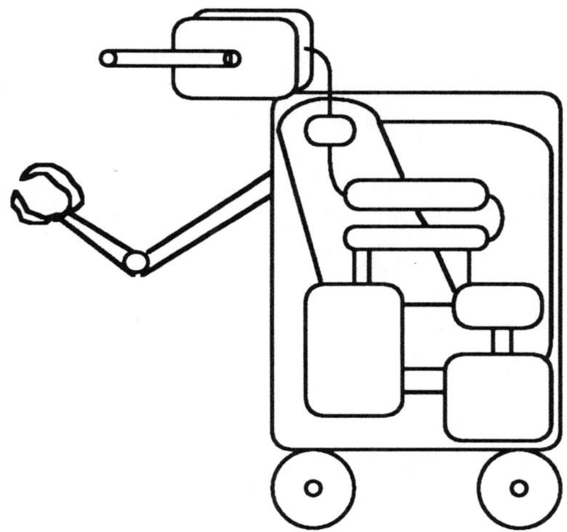

FIG. 6. 3.

The whole idea of qualia, as understood by *pholk* psychologists—that is, by philosophers who have massaged each others intuitions with a host of intuition pumps—is predicated on the mistaken idea that there is, there must be, such a central registration medium for subjectivity. Tradition has tacitly supposed that there is, but there is not. It is as simple as that.

REFERENCES

Dennett, D. C. (1991a). *Consciousness explained*. Boston: Little, Brown.

Dennett, D. C. (1991b). Lovely and suspect qualities (commentary on D. Rosenthal, The independence of consciousness and sensory quality). In E. Villanueva (Ed.), *Consciousness*. Atascadero, CA: Ridgeview.

Ornstein, R., & Thompson, R. F. (1984). *The amazing brain*. Boston: Houghton Mifflin.

Rosenthal, D. (1991). The independence of consciousness and sensory quality. In E. Villanueva (Ed.), *Consciousness* (pp. 15–36). Atascadero, CA: Ridgeview.

III MODELS OF CONSCIOUSNESS

Antti Revonsuo
University of Turku, Finland

CHASING THE GHOST IN THE MACHINE

The general strategy in cognitive modeling is to open up the big "black box" representing the organism (or its brain) and to postulate a complex system of many smaller boxes in order to depict the causal relationships prevailing between distinct mental or information-processing functions. For example, models of human memory systems have been based on such functional distinctions: memory has been divided into procedural, semantic, episodic, and working memory. The last has been further subdivided into "the phonological loop," "the visuo-spatial sketchpad," and "the central executive" (Baddeley, 1986). Jerry Fodor's (1983) ideas on the modularity of mind encouraged functionalistic thinking also in neuropsychology, where theorizing was soon started to be made in terms of "modules." Indeed, the idea of human mind as a modular system is one of the central hypotheses of modern cognitive neuropsychology. Mind and brain are believed to be so organized that cognition is mediated by large numbers of relatively independent processing modules, each of which can be separately impaired (Ellis & Young, 1988).

Functionalistic box models of cognitive processes have accumulated with increasing frenzy, but where does that leave us with consciousness? A growing fear among cognitive scientists has gradually developed into reality: Somewhere behind all those input-lexicons, shape-from-shading computations, primal sketches, and face recognition units there must be a distinctively conscious mechanism, because the modular processes are, almost by definition, reflexlike, automatic, mandatory, specific, and, above all, utterly out of the reach of conscious access. Jerry Fodor (1983) recognized this problem: All the really interesting mental processes seemed to escape a modular analysis—all that had been reasonably successfully developed was "a sort of extended psychophysics." Fodor (1983, p. 127) put the point aptly, "The ghost has been chased further back into machine, but it has not been exorcised." A similar worry has become explicit with regard to models of working memory: The "central executive" component that undoubtedly carries the weight of conscious awareness on its shoulders, has "represented an area of residual ignorance rather than a well worked out concept" (Baddeley, 1992, p. 12).

The situation we are in brings to mind echoes of the perils against which Daniel C. Dennett (1978, p. 164) warned us as early as the 1970s, "The ever-present worry is that as we devise components—lesser homunculi—to execute various relatively menial tasks near the periphery, we shall be 'making progress' only by driving into the center of our system an all-powerful executive homunculus whose duties require an almost Godlike omniscience." Dennett's forewarning seems to have been made in vain, because he recently (1991) launched a novel attack against consciousness-centered models of mind.

This reminder certainly is timely: Models of consciousness are always on the verge of slipping into infinite regresses. How can it be avoided? The problem at large is, "How do living organisms manage to interact with the world in ways that prolong their survival, optimize the production of offspring, and facilitate the persistence of their genes?" In the case of human consciousness, we would like to know in more detail how the existence of a consciously apprehended model of the world—the manifest image—aids in achieving those virtual, evolutionary goals. We must carefully tack between the rocks of postulating the explanandum too many times (which leads to regress) and the cliffs of not postulating it at all (which leads to eliminativism). Dennett's (1991) caricature of infinite regress inside the brain is the Cartesian theater, where all the wonderful experiential qualities are finally brought together in one coherent portrayal to be displayed in a 3-D, dolby-stereo, nonstop presentation for the homunculus audience. Fear of infinite regress should not, however, tempt us to commit the opposite error of claiming that there really is no phenomenology at all. How do we postulate consciousness once and only once? Perhaps the solution is, as John Searle (1992) suggested, that we should not attempt to remake the reality–appearance distinction anymore at the level of the experienced world of consciousness. The connection between world and mind necessarily involves a two-place relation: Out there is the real object (say, the three stars of the Belt of Orion) and in here is my conscious perception of it. But when we restrict our attention to consciousness itself (say, I have a dream of standing on a mountain top admiring the night sky and the Belt of Orion), we have only a one-placed relation. The phenomenology is *only experienced;* it is itself no more perceived or observed—there is no *further* distinction between the experienced world and the observer. Thus, we should not duplicate the relation between the world and the organism as the relation between the experienced world and the experiencing mind. Consciousness just consists of the experienced world, and this experiencing should not make reference to an additional observer lest we resort to committing the fatal crime of reproducing the explanandum in the explanans.

Perhaps, then, infinite regress is not inescapable after all. This must have been the conviction of those brave theorists who have dared explicitly to incorporate consciousness in their models of cognitive processing. Ray Jackendoff (1987) argued that each faculty of mind has its own characteristic chain of levels of representation, leading from the peripheral to the central levels. These chains might intersect at various points, and they do so especially at the central levels at which thought occurs, independent of sense modality. Jackendoff pointed out that consciousness is sharply distinguished by sense modality. Tactile, auditory, and visual experiences differ like night differs from day, and it is practically impossible to confuse them. Thus, he maintained, consciousness must be "projected from" such a level of representation that is still faculty specific and not independent of sense modality. This model is, nonetheless, plagued by some additional philosophical problems (for more details, see Revonsuo, chapter 11, this volume). Some theorists have attempted to unveil the neurobiological processes underlying consciousness (Crick & Koch, 1990; Damasio, 1989; Edelman, 1989). Increasing knowledge of the brain's functional neuroanatomy has revealed an utterly shattered jigsaw: There is no anatomically definable center where the world is put together for conscious perception. However, because we undeniably experience and see objects in the world as coherent and unified entities, there must be some mechanism in the brain that binds together the differentially processed features of the visual world. The search for such mechanisms is, inherently, the search for the mechanisms of consciousness.

So we undoubtedly have something to start with. But how can we tell whether a specific model of consciousness actually addresses consciousness at all—the very same phenomenon that we mean when we talk about consciousness. It is quite conceivable to have the word "consciousness" in your theory but to use it to refer to something entirely different from ordinary usage—even to such an extent that the theory ceases to be a theory of consciousness altogether. We need to have some idea of how far we can travel from our intuitive conceptions of consciousness and still continue to talk about the same thing. If the construct called consciousness in a theory of consciousness shares none of the core ideas of consciousness that we hold pretheoretically (whatever they may be), it is questionable to call it a theory of consciousness. Either we have abandoned our initial concept altogether, in which case we do not have a theory of consciousness but of something else instead, or we have preserved the original notion at least partially, in which case our theory attempts to explain consciousness (or what is left of it). But we cannot first deform the concept beyond recognition and then explain it. We have tragic examples of playing with words in such a way from other fields of life: Many so-called "democratic republics" were using the concept of "democracy" in spite of the fact that the political system itself had nothing to do with democracy as traditionally conceived. Do we have theories of consciousness that treat consciousness in the same way as Deutsche Demokratische Republic treated democracy? Of course, we can always assert that *real* democracy is not what we thought it is like but something else entirely. One may wonder whether Dennett's (chapter 2, this volume) argument is not like that, after all, when he writes that "I am not denying the reality of consciousness at all; I am just denying that consciousness, real consciousness, is like what they think it is like" (p. 62).

In the following two chapters we will take a more detailed look at two approaches to modeling consciousness. In chapter 7, Bernard J. Baars introduces the *global workspace theory* of consciousness. The basic idea behind Baars' framework is to contrast pairs of cognitive processes closely resembling each other, the only difference between them being that one of them is conscious and the other is not. He suggests that we can use such pairs to constrain our theory of consciousness that ought to explain the crucial difference between conscious and nonconscious cognitive processes in all comparable cases. Baars analyzes the differences between the distinctive processing architectures for conscious and nonconscious processes, and Baars and Newman extend this analysis, in chapter 9, by explaining how these different architectures are separately realized in the neurobiology of the brain.

Global workspace theory seems to preserve enough of our intuitions concerning consciousness: We find it quite natural to regard the mechanism the theory reveals as being in charge of consciousness—the same phenomenon that we were interested in in the first place. However, does global workspace theory escape infinite regress? In any case, it does not presuppose any anatomical center to the brain where it all comes together. But it does presuppose a functional center: Only information that enters the distributed but unique global system (and has certain additional properties) can ever become included in the contents of consciousness. This probably is, for Dennett, still a variety of the Cartesian theater, but in fact the system postulated by the theory is just an innocent theoretical entity that does not by itself entail any infinite regresses inside the brain.

In which way does this kind of model advance our understanding of consciousness? First, it differentiates the phenomenon we are interested in from other, unconscious mental phenomena. The presupposition (not shared by all) here is that there indeed is a line to be drawn between unconscious and conscious phenomena and that it is a significant line to be drawn if we ever want to understand how our minds work. Nevertheless, merely drawing a box and labeling it consciousness in one's theory of mind is not yet an explanation of what is going on *inside* the box. Next, we have to seek for the *mechanisms* that are in charge of realizing at the microlevel what we characterized as consciousness at the macrolevel of phenomena. Global workspace theory operates at both levels: It first identifies consciousness as a certain kind of mode of processing in the system and then continues to find out the mechanisms that, at least in principle, could be operating beneath the surface.

Recently, neuropsychologists have inadvertently had to face the problem of consciousness. Young and de Haan confessed that "We claim no special expertise in approaching the philosophical or psychological problems associated with the study of consciousness and awareness. The truth is that we had no more than a general acquaintance with such issues until they were forced upon us by research findings" (Young & de Haan, 1990, p. 30). Consequently, the royal neuropsychological road to awareness has been the surprising dissociation of the indirect or implicit cognitive abilities from the consciously accessible information in brain-injured patients. Schacter and his colleagues (1988) reviewed the evidence for such cases, and they tried to make sense of them in terms of what such pathologies can teach us about the normal organization of consciousness and cognition. Schacter's (1990) model contains a box explicitly labeled "conscious awareness system," thus assuming that all experiencing depends on one general mechanism and its interconnections with modular, nonconscious, special-purpose processors. The model has been later refined, to describe the role of consciousness in face processing disorders (de Haan, Bauer, & Greve, 1992).

In chapter 8, neuropsychologist Andrew Young approaches the problem of consciousness by considering these specific unexpected distortions and disconnections of awareness that occur in brain-injured patients. We meet patients who are subjectively blind but, in experimental situations, can still make surprisingly accurate visual distinctions; patients who are truly blind but insist that they are not; patients who do not recognize any faces although their information processing system seems to do so; and patients who believe that their friends and neighbors are suddenly somehow "somebody else"—the originals are believed to be replaced by facsimile robots or zombies—and even some cases who stubbornly insist on themselves being dead. Young thinks that the neuropsychological modeling of consciousness is not quite as uncomplicated as, for example, Schacter's first models suggest. However, it is possible to use the empirical findings to illuminate possible functions of consciousness, especially the role of awareness in guiding action, and the question concerning the unity of conscious experience.

Baars' and Young's accounts of consciousness are based on empirical findings but from entirely distinct sources. They disagree on one fundamental point: Is there one unitary conscious mechanism, or are there many separate ones? Baars thinks that there is ample evidence of one common, unifying system, global workspace, but Young finds the idea of a single consciousnes system implausible in the light of the neuropsychological data. Because the neuropsychological evidence does not point to the existence of a unitary conscious mechanism, Young wonders whether the subjective unity of conscious experience might be somewhat of an illusion.

Although the question is empirical and will eventually be settled only by further investigations, there are certain pretheoretical reasons to believe that an approach that postulates multiple conscious systems is less satisfactory than an explanation that refers to one common system. If each type of information had its own consciousness system, we would probably end up with an unacceptable multiplicity of different processing systems. After such an analysis, we would anyway have to ask: By virtue of which common property are all these systems capable of realizing conscious processes? How are those distributed systems of consciousness able to produce the phenomenological unity we (seem to) enjoy? Although the subjective unity of consciousness may well be merely an illusion, even the illusion must somehow be brought about by the brain. Even if we believe in a multitude of separate systems of consciousness, it appears that we have to eventually seek for some common element that explains the illusory unity. If no such common element exists, how can we ever justify picking up exactly those systems as systems of consciousness instead of some other arbitrary group?

We still have a long way to go before we can expect to have models of consciousness that cover empirical findings from all the relevant sources at the same time. Of course, we have no guarantee that such models are even possible. Anyway, the important point is that we have made a respectable start. The cognitive ghostbusting is not an altogether hopeless task. It is possible to make progress, at least as long as we are able to avoid the most enticing philosophical pitfalls. So, let's force our way into the machine and let the thrill of the chase enchant us!

REFERENCES

Baddeley, A. (1986). *Working memory*. London: Oxford University Press.

Baddeley, A. (1992). Is working memory working? The fifteenth Bartlett lecture. *The Quarterly Journal of Experimental Psychology, Section A: Human Experimental Psychology, 44A*, 1–31.

Crick, F., & Koch, C. (1990). Towards a neurobiological theory of consciousness. *Seminars in the Neurosciences, 2*, 263–275.

Damasio, A. R. (1989). Time-locked multiregional retroactivation: A systems-level proposal for the neural substrates of recall and recognition. *Cognition, 33*, 25–62.

de Haan, E. H. F., Bauer, R. M., & Greve, K. V. (1992). Behavioural and physiological evidence for covert face recognition in a prosopagnosic patient. *Cortex, 28*, 77–95.

Dennett, D. C. (1978). Toward a cognitive theory of consciousness. In D. C. Dennett (Ed.), *Brainstorms* (pp. 149–173). Brighton, Harvester.

Dennett, D. C. (1991). *Consciousness explained*. Boston: Little, Brown.

Edelman, G. (1989). *The remembered present: A biological theory of consciousness*. New York: Basic Books.

Ellis, A. W., & Young, A. W. (1988). *Human cognitive neuropsychology*. Hove & London: Lawrence Erlbaum Associates.

Fodor, J. A. (1983). *The modularity of mind*. Cambridge, MA: MIT Press.

Jackendoff, R. (1987). *Consciousness and the computational mind*. Cambridge, MA: MIT Press.

Schacter, D. L. (1990). Toward a cognitive neuropsychology of awareness: Implicit knowledge and anosagnosia. *Journal of Clinical and Experimental Neuropsychology 12(1)*, 155–178.

Schacter, D. L., McAndrews, M. P., & Moscovitch, M. (1988). Access to consciousness: Dissociations between implicit and explicit knowledge in neuropsychological syndromes. In Weiskrantz, L. (Ed.), *Thought without language* (pp. 242–278). Oxford: Oxford University Press.

Searle, J. R. (1992). *The rediscovery of the mind*. Cambridge, MA: MIT Press.

Young, A. W., & de Haan, E. H. F. (1990). Impairments of visual awareness. *Mind & Language 5(1)*, 29–48.

7 A Global Workspace Theory of Conscious Experience

Bernard J. Baars
The Wright Institute, Berkeley

Consciousness is notoriously one of the thorniest subjects in psychological science. Its study has been neglected for many years, either because it was believed to be conceptually confusing or because the relevant evidence was thought to be poor. Today, it seems that psychologists and neuroscientists are increasingly aware that we already have a great deal of evidence on conscious experience and that a theory of consciousness may be possible. Several published proposals directly address consciousness (e.g., Baars, 1983, 1988, 1992; Crick, 1984; Crick & Koch, 1990; Edelman, 1989; Newman & Baars, 1994; Norman & Shallice, 1986; Shallice, 1978), and related issues are increasingly seen as unsolvable without reference to conscious experience.

The current approach suggests a way to specify reliable empirical constraints on a theory of consciousness by contrasting well-established conscious phenomena—such as stimulus representations known to be attended, perceptual, and informative — with closely comparable unconscious ones—that are, for example, preperceptual, unattended, or habituated. By avoiding the most disputatious arguments about "boundary cases" between clearly conscious and clearly unconscious events, we can outline a large body of solid evidence, on which to build theory (Baars, 1983, 1988, 1992).

Many findings converge to show that consciousness reflects a basic architectural aspect of the nervous system, functionally equivalent to a global workspace in a parallel and distributed set of neural processors. A similarly clear approach to well-established neurophysiological findings, such as the brain structures involved in sleep and waking and neuropathology including blindsight, suggests a neurobiological interpretation of global workspace theory (Baars & Newman, chapter

9, this volume; Newman & Baars, 1994). This architectural view of conscious experience helps to clarify many difficult questions, including the role of consciousness in learning novel information, in problem-solving, voluntary control, and the like. Consciousness turns out to be an essential functional component of the nervous system, a major biological fact, playing multiple and indispensible roles in all psychological tasks. "What happens patently when an explosion or a flash of lightning startles us . . . happens latently with every sensation we receive. . . . *A process set up anywhere in the (nerve) centres reverberates everywhere, and in some way or other affects the organism throughout*" (James, 1890/1983, pp. 372, 381).

INTRODUCTION

Until recently, scientific psychologists believed that the study of consciousness was doomed to failure, because:

1. There was insufficient evidence clearly bearing on consciousness;
2. The issue was conceptually confusing (at best) and philosophically bankrupt (at worst);
3. Any attempt to construct a theory of conscious experience was hopelessly premature.

In this chapter I will suggest that these three propositions are false. In fact, a great deal of solid evidence with a direct bearing on consciousness has accumulated since the beginnings of psychophysical research in the 1840s. When examined carefully and systematically, this evidence suggests a great conceptual clarification of the issues and places great demands on any theory of consciousness, so that even today we can suggest a class of theories that seems to meet these empirical constraints and show that alternative hypotheses are only partly true or clearly ruled out (Baars, 1983, and especially Baars, 1988). Some of those excluded alternatives are discussed later. I will comment briefly on questions raised by Professors Searle and Dennett, and conclude with three vital adaptive functions played by consciousness.

WHICH EVIDENCE BEARS MOST DIRECTLY ON CONSCIOUSNESS AS SUCH?

William James suggested that "the *distribution* of consciousness shows it to be exactly such as we might expect in an organ added for the sake of steering a nervous system grown too complex to regulate itself"

(James, 1890/1983, p. 141). We can study "the distribution of consciousness" by contrasting similar pairs of information-processing phenomena, where one member of the pair is conscious and the other is not. The result is much like an experiment in which all variables are held as constant as possible, and reports of conscious experience provide the independent variable. This strategy has indeed been applied in a number of cases, even if they were not always labeled as studies of conscious experience.

Some everyday experiences can make this point. Everyone surely has the experience of repetitive stimuli fading from consciousness; the sound of a noisy refrigerator pump is a favorite example. Sometimes, when such a highly predictable train of events suddenly stops, the stimulus becomes conscious, so that we know something has changed, even though we were not aware of the train of events while it was going on. Scientific students of such habituation phenomena suggest that the nervous system must maintain a rather accurate, unconscious representation of the repetitive stimulus, because a change in any parameter of the stimulus may bring it to consciousness (Sokolov, 1963). Similarly, one may be reading a book while a friend makes some remark; one's first impulse may be to ask, "What did you say?" but suddenly, the memory of what was said comes to mind. The words must have been stored unconsciously somehow, or they would not be remembered a few seconds later. This phenomenon is one aspect of "selective attention," the fact that while we are absorbed in one train of conscious events, certain unconscious stimuli appear to be processed and represented at the same time.

Both cases raise the question, what is the difference between the same stimulus representation when it is conscious versus unconscious? In this way, we can contrast conscious contents with closely comparable mental representations that are clearly unconscious. For example, we can consider perception to be a conscious representation of sensory input. However, there are numerous examples of sensory input representations that are not conscious. Table 7.1 shows a set of contrasts related to perception that are supported by widely accepted evidence. Table 7.2 makes the same case for conscious mental imagery.

TABLE 7.1
Contrastive Evidence in Perception

Conscious Events	Comparable Unconscious Events
1. Perceived stimuli	1. Processing of stimuli lacking in intensity or duration and centrally masked stimuli
	2. Preperceptual processing
	3. Habituated or automatic stimulus processing
	4. Unaccessed meanings of ambiguous stimuli
	5. Contextual constraints on the interpretation of percepts
	6. Unattended streams of perceptual input

TABLE 7.2.
Contrastive Evidence in Imagery
Note: "Images" are broadly defined here to include all quasiperceptual events occurring in the absence of external stimulation, including inner speech and emotional feelings.

Conscious Events	Comparable Unconscious Events
1. Images retrieved and generated in all modalities	1. Unretrieved images in memory
2. New visual images	2. Automatized visual images
3. Automatic images that encounter some unexpected difficulty	3. ——
4. Inner speech: currently rehearsed words in short-term memory	4. Currently unrehearsed words in short-term memory
	5. Automatized inner speech?

TABLE 7. 3.
Contrasting Capabilities of Conscious and Unconscious Processes

Conscious Processes	***Unconscious Processes***
1. Computationally inefficient: Many errors, relatively low speed, and mutual interference between conscious processes.	1. Very efficient in routine tasks: Few errors, high speed, and little mutual interference.
2. Great range of contents.	2. Each routine process has a limited range of contents.
Great ability to relate different conscious contents to each other.	Each routine process is relatively context-free.
Great ability to relate conscious events to their unconscious contexts.	Each routine process is relatively context-free.
3. High internal consistency at any single moment, seriality over time, and limited processing capacity.	3. The set of routine, unconscious processes is diverse, can sometimes operate in parallel, and together has great processing capacity.

With regard to mental imagery, an example from Table 7.2 is easy to find. What is the difference between a visual image of this morning's breakfast before we bring it to mind and afterward? Surely the information for that image must have existed before we recalled it. Again, we have a clear contrast between two mental representations of the same event, one of which is conscious and the other not. There are dozens of such contrastive pairs of similar mental events. These are all studies of the distribution of consciousness. A complete theory of conscious experience must be able to account for all such contrasts and is therefore highly constrained.

Many more sets of contrastive pairs may be found in Baars (1988), including some involving the neurophysiological basis of conscious experience; the shaping of consciousness by unconscious "contexts"; the role of set, expectations, intentions, and voluntary control; selective attention; habituation and automaticity; and the like. A number of other researchers have used the same strategy to investigate specific phenomena (e.g., Libet, 1981; Newman & Baars, 1994; Weiskrantz, 1986).

Table 7.3 shows a set of capability contrasts for conscious and unconscious processes, describing their functional capabilities. It is currently the most theoretically important contrastive table. For our purposes, we can sum up Table 7.3 in a single question: How does a serial, integrated, and very limited stream of consciousness emerge from a nervous system that is mostly unconscious, distributed, parallel, and of enormous capacity? It is this question that suggests most strongly that consciousness corresponds to globally disseminated information in the nervous system, as we see later.

The seriality of consciousness refers to the well-known fact that conscious contents follow one after the other, like a stream. In contrast, a look at the brain suffices to show that it is extraordinarily complex, that great numbers of neurons are firing at the same time, so that single neurons, small populations of neurons, and entire neural structures operate in parallel with each other.

Again, Table 7.3 shows that the conscious capacity of immediate memory, and so forth is extraordinarily limited. On the unconscious side, processing capacity appears to be much larger. Finally, when multiple tasks are done at the same time, only a few intermittently conscious ones can be done, and even then the tasks mutually interfere and degrade in accuracy and timing. On the unconscious side, for example, just in interpreting a single sentence, we have acoustical analysis, phonological, syllabic, morphological (especially in Finnish!), lexical, syntactic, semantic, and pragmatic. All of those levels of analysis are predominantly unconscious, and they can work concurrently.

Table 7.3 also refers to the central issue of *context sensitivity* of conscious contents, where contexts are defined as unconscious knowledge structures that shape conscious experience, without themselves becoming conscious (Baars, 1988). This is well known in perception. Photographs of the moon show impact craters, largely because of the patter of light and shade; implicitly, the eye assumes that light comes from above. When we turn such a picture upside down, the craters look like hills, and vice versa.

The visual system very plausibly assumes that light comes from above—as it does, on earth. A shadow cast by a hill is therefore on the lower side of the hill. By contrast, a shadow in the upper half of a circular object normally indicates a hollow depression in the earth. But if contrary to our normal visual assumption, the light happens to come from below, the expected shadows are reversed; of course, the visual system does not "know" that and continues to interpret the scene as if light is coming from above. Therefore, turning the moon picture upside down converts concavity into convexity, and vice versa.

Contextual influences, in which conscious contents change as a result of unconscious knowledge, are utterly pervasive in perception, in mental imagery, conscious beliefs, conceptual knowledge, and the voluntary components of action. All these conscious events rest upon presuppositions of various kinds that are unconscious at the time they exercise their influence. In the current theory, context—defined as an unconscious knowledge structure that shapes conscious experience—is a major theoretical construct.

The Capability Contrasts of Table 7.3 Suggest a Global Workspace Architecture

The right half of Table 7.3 is already clear. On the unconscious side, specialized processors such as those involved in visual analysis, syntax, motor control, and face recogniton are diverse, can operate in parallel, and together have very great capacity—again, by definition of a parallel distributed system. Because no one can describe these complex processes in any detail, we consider them unconscious. As parallel "societies" of specialized processors have become more popular, global workspace architectures have been adopted to coordinate such very flexible configurations of local systems. The problem with a mere collection of specialists is that, although it can handle routine tasks just by assigning them to the appropriate specialists, it has great difficulty in adapting to new problems that may require integration of different specialties. To permit interaction between different specialists, various researchers have added a global workspace to the set of specialists, a memory whose contents are broadcast to all processors in the system (e.g., Erman & Lesser, 1975; Fehling, Altman, & Wilber, 1989; Hayes-Roth, 1985; Reddy & Newell, 1974). The global workspace is a publicity organ in the society of specialized processors. Input processors can compete for access to the global workspace, either singly or in coalition; once access is gained, the winning specialists can publicize their information to all other specialists, which can in turn interact with the global message. Figure 7.1 presents this situation schematically.[1]

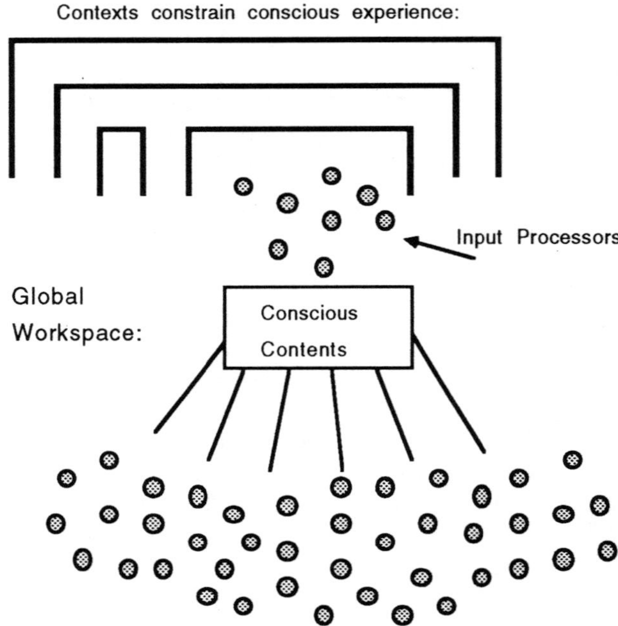

FIG 7.1. A global workspace (GW) in a distributed system.

The assumption is that the nervous system can be treated as a collection of unconscious specialized processors, including perceptual analyzers, output systems, action schemas, syntax systems, planning and control systems, and so forth. Specialized processors are recursively

[1] Dennett (1991) and Dennett and Kinsbourne (1992) provided a thorough critique of the Cartesian theater that entails the untenable notion that consciousness involves the arrival of information at some terminal, dimensionless "point." Global workspace (GW) systems superficially resemble such a Cartesian theater; however, they have been widely and successfully employed as computer architectures and do not involve an untenable notion such as the "final dimensionless point" of arrival and integration. The mere fact that GW systems have been implemented as working computer architectures shows that they cannot be Cartesian in Dennett's sense (e.g., Fehling et al., 1989; Hayes-Roth, 1985; Reddy & Newell, 1974).

organized out of other, more local processors. A syntax processor, for example, contains subsystems such as noun-phrase processors, function word processors, and so on. The system is fundamentally decentralized or distributed. Interaction, coordination, and control of the unconscious specialists requires control exercised through the global workspace. Input specialists can cooperate and compete for access to the GW; in this case, a coalition of four input specialists cooperates to place a global message. Once there, the message is broadcast to the system as a whole. The square frames above the global workspace represent contexts, defined as unconscious knowledge that constrains conscious contents without itself being conscious. Visual perception is constrained by the direction and intensity of the incoming light that is not usually conscious. Mental images are framed in a domain that resembles the visual field, but this is of course not conscious. Such context effects play a role in all conscious experiences.

An analogy may help clarify the model. Imagine a large committee of experts, enough to fill auditorium. Such an assembly would have a ready solution to any problem in the domain of any particular expert. Suppose, however, that this assembly were called upon to solve a problem that could not be handled by any expert alone. One helpful step towards solving such novel problems is to make public a global message on a large blackboard in front of the auditorium, so that in principle anyone can read the message and react. Experts could form coalitions to support certain blackboard messages, to interrupt them, change them, or decompose them into subproblems.

At any time a number of experts may be trying to write messages on the blackboard. However, if two contradictory messages were written at the same time, different coalitions of experts would be struggling against each other's messages, so that inconsistent messages would not last long. The tendency, therefore, is for consistent blackboard messages. This tendency toward internal consistency at any single time tends to serialize blackboard information, because mutually exclusive messages could only occur one after the next. And because any global message requires the cooperation, tacit or active, of many other processes, the likelihood is that novel problems would be solved more slowly than problems that fit in the domain of a particular specialist. Indeed, one function of this configuration might be to train new specialists to handle novel problems in a routine fashion. GW messages could also be used for recruitment, to elicit cooperation from other experts in the audience so that new coalitions may be established through the use of the blackboard. Finally, it can be used to coordinate coherent actions that require input from multiple systems. Global workspace configurations have numerous other uses (Baars, 1988; Fehling et al., 1989; Hayes-Roth, 1985).

This sort of situation does happen in real life. It describes fairly well the case of a legislature or committee or even a large scientific conference. Clearly, this system architecture has both advantages and disadvantages. No one is likely to use it when the problem to be solved is simple and well understood or when quick action is essential. But it is not a bad way to do things when confronted with a complex problem with no known solution, when there is time to agree or disagree, and when the costs of making an error are greater than the benefits of a quick, unreflective action.

The major constraint in this architecture is the internal consistency requirement for GW messages: From this constraint, flow all the others such as seriality (because mutually exclusive messages are competed out of existence), computational inefficiency (because all GW messages require the tacit and sometimes active inhibition of other possibilities), and limited capacity (because again, different items compete against each other, so that GW items are confined to only a single, internally consistent "chunk" at a time). These costs buy benefits, of course: the huge range of possible GW inputs; the system-wide dissemination of that information, so that any processor (or any recursive subprocessor) can be recruited if needed; and the multifarious connections that can be made between two global messages and between global and distributed sources of information (i.e., context sensitivity).

Note that GW theory has only three constructs: (a) the existence of specialized semiautonomous processors; (b) the global workspace, a memory that can be accessed by input processors and (c) contexts, the existence of processor coalitions that can influence the contents of the global workspace without themselves appearing in the GW. Figure 7.2 shows the three constructs and notes their similarities to widely accepted ideas in experimental psychology. Given the three entities, GW theory has only two processing principles: competition between GW inputs and global dissemination of output. These two principles are markers of a GW architecture and are helpful in searching for the neural correlates of consciousness (Baars & Newman, chapter 9, this volume). (Things are more complex than this. Baars (1988) suggested that conscious contents have several other necessary conditions besides global representation.)

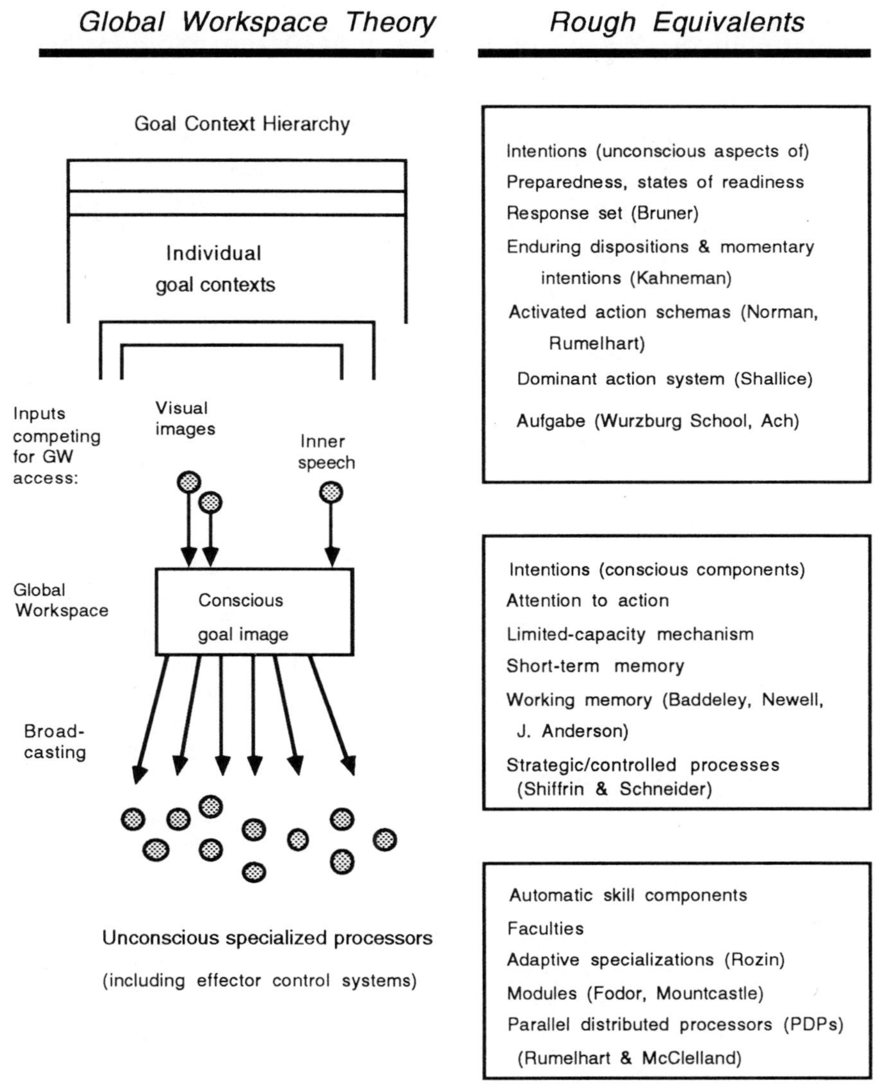

Global Workspace Theory | *Rough Equivalents*

Goal Context Hierarchy

Individual goal contexts

Intentions (unconscious aspects of)
Preparedness, states of readiness
Response set (Bruner)
Enduring dispositions & momentary intentions (Kahneman)
Activated action schemas (Norman, Rumelhart)
Dominant action system (Shallice)
Aufgabe (Wurzburg School, Ach)

Inputs competing for GW access:

Visual images

Inner speech

Global Workspace

Conscious goal image

Intentions (conscious components)
Attention to action
Limited-capacity mechanism
Short-term memory
Working memory (Baddeley, Newell, J. Anderson)
Strategic/controlled processes (Shiffrin & Schneider)

Broad-casting

Unconscious specialized processors
(including effector control systems)

Automatic skill components
Faculties
Adaptive specializations (Rozin)
Modules (Fodor, Mountcastle)
Parallel distributed processors (PDPs) (Rumelhart & McClelland)

FIG. 7.2. Rough equivalences between the basic concepts of GW theory and other widespread terms. Notice that global workspace theory has only three main constructs: the global workspace, unconscious specialized processors, and goal contexts. Each one has a graphic symbol associated with it, so that the theory can be expressed in intuitively obvious diagrams.

To many people the idea that conscious experience is used by unconscious, complex, and intelligent information processors seems counterintuitive. A good deal of the pride taken traditionally in consciousness has to do with our ability to reason consciously and explicitly, to support the distinctive and advanced aspects of human culture. But consciously deriving a geometrical proof, solving a logic problem, or thinking very clearly, step by step, about a philosophical problem, are exceptional activities for the human mind: They are culturally rare spinoffs from the everyday mundane activities of consciousness that are much more likely to leave such complex computations to highly practiced, unconscious processing routines. The whole point about distributed systems is that they require only a few crucial "hints" from conscious mechanisms; the details of motor control, or problem solving or sensorimotor integration are then solved by unconscious systems that are far more complex generally than the contents of consciousness (e.g., Greene, 1972). This is not to say that consciousness is expendable—just that its primary role is to recruit unconscious resources, to supervise, facilitate error-correction, and the like, rather than take on the burden of symbolic computation, for which it is not well-suited.

It is important to note that global dissemination of information is viewed in GW theory as a necessary but not sufficient condition for conscious experience. A more complete account must explain several other facts such as the loss of conscious access to redundant stimulation (i.e., habituation to repeated stimuli and automatization of well-practiced skills). These additional conditions for conscious experience are detailed in Baars (1988), and the interested reader is referred to that source.

The Perceptual Bias of Consciousness

It is tempting to suppose that consciousness deals mainly with "higher functions"—language, meaning, mathematics, science, and the like. Indeed, academics since Aristotle have favored this view. We argue that conscious experience has a marked preference for "plain" perception, at least when we look for the clearest and most detailed contents of consciousness. Explicit abstract reasoning, as in logic, only emerged in the last few millenia—a mere flash in the long journey of human evolution. Perhaps these sophisticated products of high culture ride on a system that is biologically specialized in perceptual functions.[2]

If we compare consciousness of perceptual input, output, and intermediate functions it becomes rapidly clear that consciousness is closely wedded to input, or the quasiperceptual input that we find in visual imagery, inner speech, anticipatory bodily feelings such as anxiety, and the like. In vision we can become conscious of whole scenes, objects within those scenes, and numerous aspects of each object—color, texture, location, reflectance, intensity, even emotional tone, reminders of similar objects experienced in the past, and so forth. Internally generated quasiperceptual events—images, episodic memories, problem solving such as mental arithmetic, word-retrieval, and so on—are less richly detailed, compared to conscious perception.

When we examine abstract concepts and beliefs or the interplay of our own expectations and intentions, detailed conscious contents become increasingly spare. Finally, when we try to experience action control or the detailed workings of a skilled act such as reading, we seem to lose even more detailed conscious access. Sometimes we are simply wrong about the details of action. For instance, in reading, most people intuitively think that their eyes are scanning the words continuously. One need only observe another person reading to realize that this commonsense belief is utterly false: The eyes can plainly be seen to jump from point to point most of the time. Similarly, most people have no idea about the extraordinary speed and precision of their own vocal tracts—for example, the delicate millisecond delay between opening the lips and voice onset in pronouncing the syllable *ba* (Liberman, Cooper, Shankweiler, & Studdert-Kennedy, 1967).

Of course, conscious perceptual feedback from motor activities is crucial in controlling the loudness of one's voice, in correcting errors, and in shaping actions in general. When people listen to loud music on headphones, they tend to speak very loudly; without conscious feedback, they lose control of speech amplitude. All motor control relies on feedback, but by this we mean perceptual feedback (Levelt, 1989); and now we find ourselves back in the perceptual system. Thus, consciousness in general seems to have a marked affinity for perceptual processes compared to skilled actions, abstract concepts, or other internal functions.

What does this perceptual affinity imply for theories of consciousness? For global workspace theory, it suggests that perceptual

[2] There are significant exceptions to this rough generalization, namely the cases of intentions and expectations, sometimes called "nonqualitative" conscious contents, as well as conscious access to abstract (nonimageable) concepts. These questions are discussed in detail in Baars (1988).

systems may provide the most important input processors to the biological global workspace.

Further Contrastive Evidence Constrains More Advanced Models

Additional contrastive data-sets are explored in *A Cognitive Theory of Consciousness* (Baars, 1988), leading to increasingly adequate and inclusive GW models. Very briefly, these models appear to explain:

1. the important role of unconscious contexts, knowledge that shapes conscious experience without itself becoming conscious;
2. loss of conscious access in habituation and automaticity;
3. the role of conscious and unconscious processes in "spontaneous" (inexplicit) problem solving;
4. the intimate connection between consciousness and control of voluntary action (Baars, 1992);
5. the relationship of consciousness to selective attention;
6. how the workings of consciousness may lead to other limited-capacity phenomena such as short-term memory;
7. the relationship of consciousness to a "self-system" that can also be defined by contrastive analysis;
8. the numerous adaptive roles consciousness seems to perform in the nervous system.

I have been repeatedly surprised by the fit of GW theory with new domains of evidence for which the original theory was not designed. For a recent example, we can now hazard a reasonable hypothesis of the neural basis of conscious experience (see Baars & Newman, chapter 9, this volume). This comes as a pleasant surprise, because GW theory was originally developed only to deal with psychological evidence.

Are We *Really* Talking About Consciousness?

Professor John Searle (in discussion) asked whether we are truly talking about consciousness in this kind of approach? I would say yes. In part, this is just the standard scientific answer to the same kind of question about theories of planetary motion, for instance. Observers since ancient times have suggested that the planets were wandering, far-away bodies in the night sky, and with Kepler and Newton, those complex motions assumed a new and previously unsuspected simplicity. To the question, "But are you sure this *is* planetary motion?" Newton might have answered, "Yes, because that's what we have been observing, after all."

There is a circular quality to this traditional scientific answer; sometimes the nature of what is observed, especially when it is inferential (such as the atom in 19th century physics) can be debated for a hundred years before scientific consensus emerges. Nevertheless, once the evidence is believed to be understandable in terms of atoms, that is what the theory is about.

There is another answer that is unique to psychology: that is an appeal to personal experience. GW theory aims to understand widely shared reports of private experience. We know of considerable individual variation in such matters as visual imagery and even inner speech. But I do not know of anyone who does not report loss of conscious access to redundant stimulation, to highly practiced actions, to perceptual experience as it fades into memory, and the like. If such exceptional people exist, they would be fascinating to study, just as split-brain patients are exceptionally interesting and important, though we cannot generalize, in a superficial way, from them to the rest of humanity.

Appeal to personal experience can be viewed as unnecessary. GW theory can be constructed perfectly well on intentional reports of private experience. Strictly speaking that is true. But it is naive to think that our private experience has no persuasive force in the matter.

We noted previously that access to a global workspace and broadcasting of conscious contents are necessary but not sufficient for conscious experience in GW theory. Other necessary conditions include informativeness, internal consistency, probably some minimum duration (about 100 ms? See Blumenthal, 1977; Libet, 1981), and access by a self-system (Baars, 1988, pp. 362–363, and corresponding chapters). We may need to restrict GW input to perceptual or imagistic information. There may be other conditions we do not know about.

Access by a self-system is particularly interesting and may go some way toward explaining the extraordinary importance of conscious experience to each of us. Consciousness is our access to reality and in turn shapes us in every way (Dennett, 1978). In GW theory, the self-system is viewed as the long-lasting components of the context hierarchy, involving our most fundamental goals and our most basic expectations about the world. It is not conscious in detail, but aspects of it can become conscious upon violation. This way of thinking about the first-person perspective is psychologically productive. Of course, thinking about it is inherently third person, whereas being it is first person. In this effort, the aim of psychological theory is not to answer the question, "What is it like to be a bat?" (Nagel, 1974), but to make sense of the third-person question, "Can we understand the role of consciousness in the human nervous system?" If we ever find out what it is like to be a bat, that will be pure icing on the cake.

Finally, we should note the epiphenomenalist argument that all these matters can be talked about in a pure, physicalistically untainted third-person vocabulary of information processing or brain events. We could avoid the dreaded word consciousness by using a euphemism. One answer suggested by Professor Searle is that we would be throwing away a perfectly useful source of evidence, something any scientist should think twice about. Another is that we would be engaging in an empty ritual of self-alienation in pursuit of some ideal of physicalistic perfection. Good science is done much more pragmatically.

RELATIONSHIP TO OTHER THEORIES

Contrastive analysis is used to constrain theories of consciousness, so that one set (GW models) seems to fit, whereas other candidates may be ruled out or found to be incomplete. Which theories and hypotheses appear to be ruled out by the evidence? Here are some examples.

Activation Threshold Theories. As early as 1824, Johann Herbart wrote, "As it is customary to speak of an entry of ideas into consciousness, so I call the threshold of consciousness that boundary which an idea appears to cross as it passes from the totally inhibited state into some . . . degree of actual (conscious) ideation" (Herbart, 1824/1961, p. 40). This is a perfectly good idea for modeling the absolute psychophysical threshold and for convergent priming effects in memory (Anderson, 1983). But it leads to a fundamental contradiction when we consider automaticity, the loss of conscious access to well-practiced skills. The more practice we get in some predictable task, the more easily it is performed, but the less we are conscious of it, and the less we can report accurately on its details. Automaticity can develop quite rapidly in simple tasks (e.g., Langer & Imber, 1979). Greater ease of processing, improvement in response times, a drop in errors, and so on can be explained by increased activation; but because improved skill goes along with a loss of conscious access, the increase in activation cannot also model access to consciousness. This is a paradox.

One way to resolve the paradox is to have two different kinds of activation. MacKay (1987), for example, proposed the term priming versus activation to deal with this issue. Priming leads to conscious access, whereas activation increases skill. But this is only a local fix, reminiscent of Freud's use of two kinds of neurons, namely conscious ones (called *psy*) and unconscious ones (*phi*) (Freud, 1895/1966). It is a description, not an explanation.

The simplest GW treatment would suggest that MacKay's priming is the probability of access to the global workspace (or its functional equivalent), whereas activation has to do with the recruitment of a new processing coalition, a new specialized processor, whose sole task is to perform the practiced skill. The more this novel coalition becomes modularized, the less it requires GW access, and the more rapid, accurate, and resistant to interference the skill becomes. This goes beyond a mere labeling differentiation between priming and activation, and thereby breaks the circularity of the descriptive solution.

The activation paradox and its solution within a richer theory points to the importance of considering multiple bodies of evidence, rather than a single phenomenon such as memory activation, psychophysical threshold, or automaticity. Contrastive analysis works best when simultaneous, multiple contrasts are used to constrain the theory. Unfortunately, until recently this kind of domain integration was not widely taught in empirical psychology. Psychological phenomena are usually so complicated and subtle that we usually have our hands full just learning more and more about any single domain of evidence. I would hope for more tolerance and understanding of integrative theory. Resarch in single domains is indispensible, of course; but fitting the pieces together into a larger puzzle is also necessary to healthy scientific development.

Unified Theories. Alan Newell has emphasized the need to integrate multiple bodies of evidence in a unified psychological theory (Newell, 1990). Unfortunately, Newell's sophisticated SOAR theory has no stated role for consciousness at all. Although SOAR has a working memory, it is one with indefinitely large capacity, and all the automatic condition-action rules (productions) that resemble parallel distributed systems place their "actions" in working memory as soon as it displays the appropriate conditions. SOAR has no complex and sophisticated specialized processors in the sense used in GW and other theories. The working memory component takes on the burden of every single production that is triggered, so that the more is done automatically, the more working memory must hold. This is highly implausible. It fails to deal adequately with limited capacity mechanisms, one of the major architectural features of human psychology (Baddeley, 1992; LaPolla & Baars, 1992; Newell & Simon, 1972).

Obviously GW theory has not been tested on all domains of evidence, including some that are handled well by SOAR. It would be foolish to claim that GW theory is better in all respects than other architectural theories. Currently, it is a better way to understand the conditions of conscious versus unconscious functioning; but because this is a major architectural aspect of human functioning, this is no trivial difference. In future development, GW theory may well converge with other integrative theories.

Connectionism. The most prominent development in recent theoretical psychology is the rise of connectionist models (Rumelhart, McClelland, & the PDP Reseacrh Group, 1986). Parallel distributed processes (PDPs) perform many sophisticated functions, like the specialized processors suggested by GW theory. PDP modelers have produced many of the domain-specific models that allow us to simply refer to a class of specialized processors without stating them in detail. No architectural theory could be taken seriously in the absence of a sizable body of domain-specific models. It is also noteworthy that McClelland (1986) presented a connectionist GW system. This is a promising avenue for development.

Baddeley's Model of Working Memory. The burden of this section is to show how different models and hypotheses can be constrained by contrastive analysis. Recent empirical results in Baddeley's well-established research on working memory suggest a need for a model of conscious and unconscious mental processes to understand short-term memory functions. The working memory model suggested by Baddeley (1986) has two main components: a phonological (rehearsal) loop for verbal items and a visuospatial sketchpad for visualizable information. This straightforward model can accomodate a great deal of evidence. However, this conception of working memory assumes that the two main components function independently, but this assumption has been challenged by recent data, indicating that the phonological loop interfers with the visuospatial sketchpad, precisely as we would expect from GW theory. A single conscious stream of events may underlie both components—a deeper layer of organization on which working memory operates (Baars, 1988; Baddeley, 1992).

All these points indicate that GW theory is specific enough to exclude or suggest improvements in other models, as it indeed must to be viable. No doubt the theory can be made more detailed and explicit than it is today. This is a major aim for the next decade.

Consciousness Versus Selective Attention

A useful distinction, implicit in general usage, can be made between consciousness and attention. We are conscious of a cup, a sound, a visual image, a feeling, a thought. But we pay attention to a source of information when there is a choice between different sources (James, 1890/1983). Once we pay attention to the printed word in front of us, rather than to some inner speech or feelings or the music in the background, we become conscious of whatever we have chosen. Thus, in general usage, attention involves a selection (voluntary or automatic) between possible conscious contents. To study attention, we study the process of selection. To study consciousness as such, we need to compare a conscious experience of a cup to the unconscious representation of the cup after habituation, preperceptually (e.g., Libet, 1981) or after a blindsight lesion to the striate cortex (see previous discussion).

Selective attention studies since the 1950s have focused on the presumed "filtering" process that allows one to choose one rather than another stream of input. But those studies have not often been used to cast light on consciousness as such, though recently, of course, that has become a much more legitimate topic. Consciousness and attention are, of course, closely allied topics, but they can be studied separately (e.g., Baars, 1988).

Not All Theaters Are Cartesian

Dennett (1991) and Dennett and Kinsbourne (1992) have had widespread impact with a critique of the Cartesian theater, a fallacy they consider to be part of our commonsense view of consciousness. The Cartesian theory is a point where it all comes together, where conscious perceptual input is determined and from which voluntary action is planned and executed. Of course, a dimensionless point where input and output converge is absurd by modern scientific standards, just as Descartes' choice of the pineal gland (as the only nondual structure in the brain and therefore the logical point for the soul to interact with the body) is absurd. In that sense, Cartesian theater is a well-chosen label.

But not all theaters are Cartesian. A real theater with a stage on which plays are performed is non-Cartesian. An army headquarters where incoming information is displayed and general orders are issued is likewise non-Cartesian. Finally, a GW architecture with a memory that is broadcast to multiple specialized processors somewhat like the stage of a theater is non-Cartesian. Dennett and Kinsbourne performed a service in alerting us to a fallacious theater metaphor, but their critique does not apply to theater metaphors that are workable, useful, and actually implemented by cognitive scientists (e.g., Fehling et al., 1989; Hayes-Roth, 1985).

What Is Missing in the Searchlight Hypothesis

Some proposals today were developed specifically to deal with conscious experience such as the searchlight hypothesis of Crick (1984) and the notion of a "binding frequency" for unifying visual components into a coherent conscious scene (Crick & Koch, 1990). Kinsbourne advanced a notion of "cortical focus" (1993), whereas Edelman (1989) developed the concept of a "re-entrant" loop corresponding to focal consciousness. Although these proposals are not psychologically as explicit as the current theory, they do make testable predictions about neural mechanisms (see Baars & Newman, chapter 9, this volume).

Scientific metaphors carry their own pros and cons. The Rutherford atom of the late 19th century compared the atom to a tiny solar system—a very productive way of looking at the massive nucleus but a profoundly misleading way to view electrons. From the perspective of GW theory, the problem of a searchlight metaphor is that (a) it has not been implemented in nearly enough detail to simulate the effects it is supposed to have, unlike the GW systems that have been extensively tested in practical applications; and (b) the searchlight metaphor leaves a great deal unexplicated. One thing that not explicitly worked out in the searchlight concept is the reason why the searchlight should be targeted at a specific part of cortex, for example. In addition to targets, real searchlights have audiences and directors; there are humans who aim the searchlight to achieve the goal of conveying information to an audience. Without a directing intelligence, a governing purpose, and an audience, a real searchlight is useless. Yet these aspects of a searchlight metaphor have not been articulated. If these implicit features were worked out in detail, the searchlight metaphor might be difficult to distinguish from a GW architecture.

THE ADAPTIVE FUNCTIONS OF CONSCIOUSNESS

A fuller development of GW theory indicates that consciousness performs at least 10 distinct adaptive functions (Baars, 1988). The following are examples.

The Adaptation and Learning Functions of Consciousness. Conscious experience points to novel events, so that multiple unconscious processors can go to work on those events—it is ostensive. In many visual puzzles, for example, we need only look at the anomalous object, and the unconscious intelligence of the visual system takes it from there. This is especially shown by such phenomena as perceptual adaptation to profoundly transformed visual input by means of prism goggles. The same point is made in recent research especially by implicit learning, in which the rules learned never become conscious, though the events that give rise to rule-learning must be conscious for learning to occur (e.g., Reber, 1992).

The relationship between learning and consciousness is often posed in an all-or-none fashion, whether learning is possible without consciousness. This way of posing the question makes conclusive answers hard to find, simply because we are not in a good position currently to be sure that learning occurs without any conscious involvement whatsoever (Holender, 1986). It is much easier to show that the more information we need to learn, the longer we must be conscious of the material to be learned. In learning research the length of conscious exposure is typically called "number of trials," "rehearsal time," and so forth. But of course, it really means how long the subjects are asked to be conscious of the material.

Editing, Flagging, and Debugging Functions of Consciousness. A number of cognitive psychologists have suggested that consciousness becomes especially important when there are discontinuities, violations of expectations, or errors (e.g., Mandler, 1984). Here again, consciousness operates as a pointing device, an ostensive function. We are very rarely able to state a speech error consciously in terms of the syntactic rules violated, the phonology, and so on. We simply know that it is an error, and with a moment of thought we can produce a correction. The detailed intelligence behind these abilities is unconscious. Unconscious systems detect the error (this is part of the editing function); then we can "flag" the error by becoming conscious of it; and finally, unconscious systems "debug" the error, allowing a correction to become conscious.

Recruiting and Control Functions of Consciousness. Conscious goal images can recruit subgoals and motor systems to organize and carry out actions. By broadcasting a goal image, the conscious component of the nervous system can trigger specialized unconscious processors that are able to accomplish the goal. In related work, a modern version of William James "ideomotor theory of action control" can be shown to fit the GW architecture (e.g., Baars, 1987, 1988).

These are only some of the adaptive functions that seem to require conscious involvement. There are at least a half dozen more (Baars, 1988). Consciousness turns out to be anything but some illusory epiphenomenon. This is reassuring, because no epiphenomenon has been found in nature so far!

ACKNOWLEDGMENT

I am most grateful for Dr. Katharine A. McGovern for helpful comments in the presentation and writing of this chapter.

REFERENCES

Anderson, J. R. (1983). *The architecture of cognition.* Cambridge, MA: Harvard University Press.

Baars, B. J. (1983). Conscious contents provide the nervous system with coherent, global information. In R. Davidson, G. Schwartz, & D. Shapiro (Eds.), *Consciousness and self-regulation* (pp. 45–76). New York: Plenum.

Baars, B. J. (1987). What is conscious in the control of action? A modern ideomotor theory of voluntary action. In D. Gorfein & R. R. Hoffman (Eds.), *Learning and memory: The Ebbinghaus centennial symposium* (pp. 209–236). Hillsdale, NJ: Lawrence Erlbaum Associates.

Baars, B. J. (1988). *A cognitive theory of consciousness.* New York: Cambridge University Press.

Baars, B. J. (1992). *Experimental slips and human error: Exploring the architecture of volition*. New York: Plenum.

Baddeley, A. D. (1986). *Working memory.* Oxford: Clarendon Press.

Baddeley, A. D. (1992). Consciousness and working memory. *Consciousness & Cognition, 1(1),* 3–6.

Blumenthal, A. L. (1977). *The process of cognition.* Englewood Cliffs, NJ: Prentice-Hall.

Crick, F. H. C. (1984). Function of the thalamic reticular complex: The searchlight hypothesis. *Proceedings of the National Academy of Sciences, 81,*4586–4590.

Crick, F. H. C., & Koch, C. (1990). Towards a neurobiological theory of consciousness. *Seminars in Neurosciences, 2,* 263–275.

Dennett, D. C. (1978). Toward a cognitive theory of consciousness. In D. C. Dennett (Ed.), *Brainstorms* (pp. 149–173). New York: Bradford Books.

Dennett, D. C. (1991). *Consciousness explained.* Boston: Little, Brown.

Dennett, D., & Kinsbourne, M. (1992). Time and the observer: The where and when of consciousness in the brain. *Brain and Behavioral Sciences, 15(2),* 183–247.

Edelman, G. (1989). *The remembered present: A biological theory of consciousness*. New York: Basic Books.

Erman, L. D., & Lesser, V. R. (1975). A multi-level organization for problem-solving using many, diverse, cooperating sources of knowledge. In *Proceedings of the 4th Annual Joint Computer Conference Georgia, USSR,* (pp. 87–93).

Fehling, M. R., Altman, A., & Wilber, B. M. (1989). The heuristic control virtual machine: An implementation of the schemer computational model of reflective, real-time problem-solving. In V. Jaggannathan, R. Dodhiawala, & L. Baum (Eds.), *Blackboard architecures and applications* (pp. 37–59). Boston: Academic Press.

Freud, S. (1966). Project for a scientific psychology. In J. Strachey (Ed.), *The standard edition of the complete psychological works of Sigmund Freud, Vol. 3* (pp. 35–77). London: Hogarth Press. (Original work published 1895)

Greene, P. H. (1972). Problems of organization of motor systems. *Journal of Theoretical Biology, 3,* 303–308.

Hayes-Roth, B. (1985). A blackboard architecture of control. *Artificial Intelligence, 26,* 251–351.

Herbart, J. (1961). Psychology as a science, newly founded upon experience, metaphysics and mathematics. In T. Shipley (Ed.), *Classics in Psychology* (pp. 921–947). New

York: Philosophical Library. (Original work published 1824)

Holender, D. (1986). Semantic activation without conscious identification in dichotic listening, parafoveal vision, and visual masking: A survey and appraisal. *Behavioral and Brain Sciences, 9*, 1–66.

James, W. (1983). *The principles of psychology*. Cambridge, MA: Harvard University Press. (Original work published 1890)

Kinsbourne, M. (1993). Integrated field model of consciousness. In G. R. Brock & J. Marsh (Eds.), *CIBA symposium on experimental and theoretical studies of consciousness* (pp. 51–60). London: Wiley.

Langer, E. J., & Imber, L. G. (1979). When practice makes imperfect: The debilitating effects of overlearning. *Journal of Personality and Social Psychology, 37(11)*, 2014–2024.

LaPolla, M., & Baars, B. J. (1992). An implausible cognitive architecture. Commentary on A. Newell, SOAR as a unified theory of cognition: Issues and explanations. *Behavioral and Brain Sciences, 15(3)*, 464–492.

Levelt, W. J. M. (1989). *Speaking: From intention to articulation*. Cambridge, MA: MIT Press.

Liberman, A. M., Cooper, F., Shankweiler, D., & Studdert-Kennedy, M. (1967). Perception of the speech code. *Psychological Review, 74*, 431–459.

Libet, B. (1981). Timing of cerebral processes relative to concomitant conscious experiences. In G. Adam, I. Meszaros, & E. I. Banyai (Eds.), *Advances in physiological science, Vol. 17* (pp. 313–317). Elmsford, NY: Pergamon.

MacKay, D. G. (1987). *The organization of perception and action*. New York: Springer.

Mandler, G. (1984). *Mind and body: Psychology of emotion and stress*. New York: Norton.

McClelland, J. L. (1986). The programmable blackboard model of reading. In J. McClelland, D. E. Rumelhart, & the PDP Research Group (Eds.), *Parallel distributed processes: Explorations in the microstructure of cognition, Vol. 2* (pp. 122–169). Cambridge, MA: MIT Press.

Nagel, T. (1974). What is it like to be a bat? *The Philosophical Review, 83*, 435–450.

Newell, A. (1990). *Unified theories of cognition*. Cambridge, MA: Harvard University Press.

Newell, A., & Simon, H. A. (1972). *Human problem-solving*. Englewood Cliff, NJ: Prentice Hall.

Newman, J., & Baars, B. J. (1994). A neural attentional model for access to consciousness: A global workspace perspective. *Concepts in Neuroscience, 2(3)*.

Norman, D. A., & Shallice, T. (1986). Attention to action: Willed and automatic control of behavior. In R. Davidson, G. E. Schwartz, & D. Shapiro (Eds.), *Consciousness and self-regulation, Vol. 4* (pp. 63–88). New York: Plenum.

Reber, A. S. (1992). The cognitive unconscious: An evolutionary perspective. *Consciousness & Cognition, 1(2)*, 93–133.

Reddy, R., & Newell, A. (1974). Knowledge and its representations in a speech understanding system. In L. W. Gregg (Ed.), *Knowledge and cognition* (pp. 30–59). Potomac, MD: Lawrence Erlbaum Associates.

Rumelhart, D. E., McClelland, J., & the PDP Research Group (1986). *Parallel distributed processes: Explorations in the microstructure of cognition, Vol. 1*. Cambridige, MA: MIT Press.

Shallice, T. (1978). The dominant action system: An information processing approach to consciousness. In K. S. Pope & J. L. Singer (Eds.), *The stream of consciousness: Scientific investigation of the flow of experience* (pp. 97–135). New York: Plenum.

Sokolov, E. N. (1963). Perception and the conditioned reflex. New York: MacMillan.

Weiskrantz, L. (1986). *Blindsight: A case study and implications*. Oxford: Clarendon Press.

8 Neuropsychology of Awareness

Andrew W. Young
MRC Applied Psychology Unit, England

FORMS OF AWARENESS

In the 19th century, the study of consciousness and the relation between conscious and unconscious processes was considered central to the fledgling disciplines of neurology, psychology and psychiatry (Ellenberger, 1970), yet for much of the 20th century it has been deliberately marginalized for what many people still consider to be good reasons. One therefore approaches the issues involved with considerable trepidation, but there can be no doubting that they have resurfaced during the last 25 years, and especially in neuropsychology.

The principal reason for this renewed interest in the neuropsychology of consciousness is empirical; it has become clear that loss of different aspects of awareness can form a central feature of certain types of neuropsychological impairment. Many neuropsychologists now accept that we cannot simply ignore these striking phenomena, and are intrigued by the possibility of an empirically-based approach to understanding consciousness and awareness. Here, I will summarize some of the progress that has been made, and explore a few implications of this work.

Discussions of consciousness, awareness, and related mental phenomena often assume that everyone knows what they are talking about, and that they are talking about the same things. This has rightly been considered unwise (Allport, 1988), so I will begin by distinguishing four different senses of awareness:

1. *Phenomenal awareness*: the experience of seeing, hearing, touching, and so on.
2. *Access awareness*: where stored information is brought to mind, such as when recognizing an object or a face, remembering something that happened in the past, and so on.
3. *Monitoring*: including awareness of our own actions and their effects and monitoring perceptual information for discrepancies with current plans and hypotheses.
4. *Executive awareness*: awareness of our goals and intentions.

This list, which is grounded in the distinctions made by Block (1991), is not meant to be exhaustive; merely convenient for present purposes. Neither do I want to imply that there are not different forms of awareness within each of these different categories; it seems obvious enough that the phenomenal awareness involved in seeing something is not like phenomenal awareness of hearing or touch and that even within vision the phenomenal "feel" of color, shape, and other visual attributes can be quite different. So any satisfactory account of awareness is probably going to look quite complicated, and we may need different types of account for these different senses of awareness.

A wide range of neuropsychological impairments can be considered relevant to understanding awareness (Schacter, McAndrews, & Moscovitch, 1988; Young & de Haan, 1990). I use five initial examples: visual field defects, achromatopsia, prosopagnosia, amnesia, and anosognosia. I have chosen these because they can be fitted reasonably neatly into the crude typology implicit in the list of different senses of awareness given previously. In each case, one can argue that there are relatively circumscribed deficits that primarily impinge on one of these forms of awareness.

VISUAL FIELD DEFECTS

These form by far the most widely known example, because of the seminal work on blindsight (Pöppel, Held, & Frost, 1973; Weiskrantz, 1980, 1986, 1987, 1990; Weiskrantz, Warrington, Sanders, & Marshall, 1974). There is compelling evidence of accurate responses to stimuli that people insist that they do not "see", and there has been considerable progress in identifying the visual pathways that mediate these effects (Cowey & Stoerig, 1991, 1992).

One of the most commonly observed effects of brain injury is loss of vision for an area of the visual field. Often this takes the form of a hemianopia (loss of half the field of vision), but sometimes the area of loss of vision (the scotoma) is smaller than this. To test for a visual field defect the person is usually asked to report what he sees when stimuli are presented at different locations in the visual field. The point of interest here is that phenomenal awareness is lost for stimuli presented within the scotoma, yet in cases of blindsight, some forms of visual response can be demonstrated.

This has been very thoroughly demonstrated for DB, a person studied by Weiskrantz and his colleagues, who underwent an operation that necessitated removal of striate cortex from the right cerebral hemisphere. Because the optic nerves project to striate cortex, this operation left DB with a substantial scotoma. Initially, the perimetrically blind area occupied almost the entire left half of his field of vision, but over the next few years, it gradually contracted until only the lower left quadrant was involved.

Despite the scotoma, accurate responses were obtained across a wide range of stimulus eccentricities when DB was asked to point to where he guessed a flash of light had been presented. Even at his most accurate, though, performance for DB's blind field was somewhat below what he could achieve using the part of his field that still had normal vision.

Weiskrantz and his colleagues (1974) also showed that DB could discriminate the orientation of stimuli presented within his scotoma, and by asking him to guess whether a presented stimulus was a sine-wave grating of vertical dark and light bars, they were able to determine his visual acuity (in terms of the narrowest grating that could be detected). Gratings with bar widths of 1.5' could be detected in the sighted part of his field of vision, and a rather less fine but still impressive 1.9' in the "blind" field.

Later studies have confirmed and extended the original findings. In particular, DB could detect the presence or absence of a light stimulus even when it was introduced or extinguished quite slowly; he could readily distinguish static from moving stimuli; and more detailed testing of acuity in the scotoma showed that (unlike normal vision) it increased as the stimuli were moved to positions further away from fixation.

Because patients with blindsight insist that they do not see stimuli presented in the scotoma, the studies discussed so far rely on forced-choice paradigms in which they are asked to make responses that are (subjectively) guesses. These guesses can be remarkably accurate, but there are some problems associated with the technique (Weiskrantz, 1980, 1986, 1990). Not all patients are willing to engage in guessing, and there are problems in determining the criteria and strategies used in responding. Even those patients who become very adept at guessing can need to learn what to attend to when they make their guesses, though they may remain insistent that this is nothing like a visual experience.

Because these technical problems can affect studies using forced-choice guessing, it is useful to know that blindsight phenomena can be demonstrated by very different methods; Weiskrantz (1986, 1990) gave authoritative reviews. For example, it is possible to demonstrate reflex reactions in the form of skin conductance changes (Zihl, Tretter, & Singer, 1980) or altered pupil diameter (Weiskrantz, 1990). Even more intriguingly, several studies have demonstrated interactions between stimuli presented in the blind and sighted parts of the visual field. Pizzamiglio, Antonucci, and Francia (1984) noted that a full-field rotating disk produced a larger subjective tilt than did stimulation of one visual hemifield alone, for both normal and hemianopic subjects. Marzi, Tassinari, Agliotti, and Lutzemberger (1986) found that bilateral light flashes produced faster reaction times to detect a flash than did a single unilateral stimulus, even when one of the bilateral flashes fell in a hemianopic area of the visual field, from which a single flash would not have been detected. Studies of blindsight, then, have shown that the processing of visual stimuli can take place even when there is no phenomenal awareness of seeing them.

ACHROMATOPSIA

Achromatopsic patients experience the world in shades of gray, or in less severe cases, colors can look very washed out (Meadows, 1974b). This is quite different to the forms of colorblindness produced by deficiencies in one of the three types of cone receptor in the retina, for which there is still experience of color, but certain colors are not discriminated from each other. Instead, severe achromatopsias produced by cortical injury are described by the patients themselves in terms that suggest they are experienced as more like watching black and white television.

The full details of the mechanisms underlying human color vision are not known, but some of their essential features have been established. The retina contains three types of cone, each of which is maximally sensitive to light of a different wavelength. Outputs from these three types of cone are converted into color-opponent signals for red-green and blue-yellow dimensions and a separate luminance response.

This arrangement means that people are more sensitive to light of certain wavelengths than others, and the spectral sensitivity function for normal daytime (photopic) vision has distinct sensitivity peaks that because of the opponent cone mechanisms, do not correspond to the absorbence peaks of the cones themselves. This can be seen in the results for a normal observer, AC, shown in Fig. 8.1, where there is a clear increase in sensitivity for stimuli at 450, 525, and 600 nm, in comparison to the adjacent wavelengths tested (Heywood, Cowey, & Newcombe, 1991).

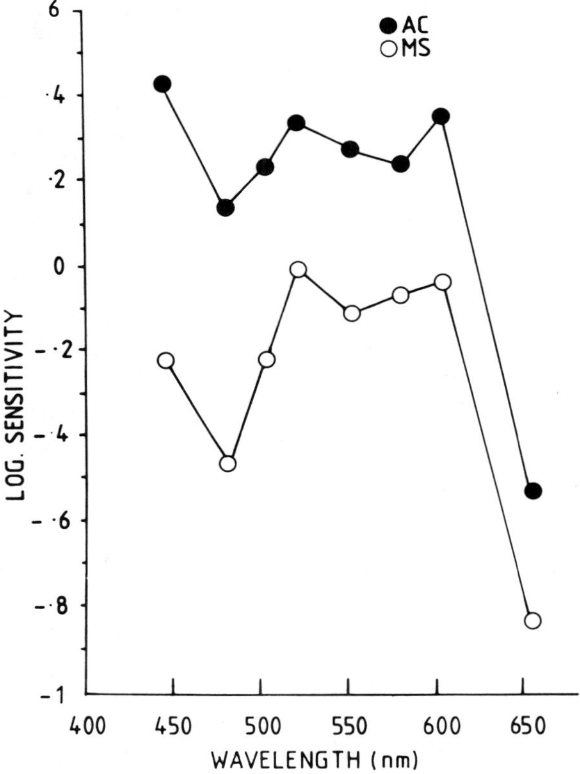

FIG 8.1. Photopic spectral sensitivity of normal observer AC and achromatopsic patient MS. Reproduced by permission of Oxford University Press from Heywood, C. A., Cowey, A., & Newcombe, F. (1991). Chromatic discrimination in a cortically blind observer. *European Journal of Neuroscience, 3*, 802–812.

Figure 8.1 also shows results for an achromatopsic patient, MS, who has no experience of color (Heywood et al., 1991; Mollon, Newcombe, Polden, & Ratcliff, 1980; Newcombe & Ratcliff, 1974; Ratcliff & Newcombe, 1982). In this test, MS described all the stimuli he saw as "dim white" or "gray", despite the differences in wavelength. Similarly, other work established that MS could not match or name colors and performed at random on the Farnsworth-Munsell 100 hue test (total error score = 1245). However, Mollon, et al. (1980) showed that he retained three functional cone mechanisms, using Stiles's two-color increment threshold procedure (Stiles, 1978).

As Fig. 8.1 shows, MS's sensitivity was poorer than AC's at all wavelengths tested. More importantly, though, his spectral sensitivity showed the same peaks as AC's, demonstrating the presence of opponent cone mechanisms despite the complete loss of color experience. This is a different result than would be expected either from monochromatic vision due to loss of any two types of cone or for vision based only on the retinal rod receptors. Interestingly, measurements of spectral sensitivity with a forced-choice guessing response can also be used to demonstrate opponent cone mechanisms in the scotomas of patients with blindsight (Stoerig & Cowey, 1991).

In achromatopsia, the subjectively described experience of seeing things in shades of gray suggests that patients should remain relatively well able to distinguish brightness differences, and this was certainly true for MS (Heywood et al., 1991). For example, he could find the odd gray in an array of nine gray squares for which eight were of identical brightness (72% correct; chance performance would be 11% correct). In contrast, when MS was asked to find the colored square among an array containing eight other squares that were all grays of different brightness, he performed at chance level (10% correct), which confirms the absence of any subjective experience of color. Even when presented with a vertical row of three separate isoluminant red or green colored squares and asked only whether it was the top or the bottom square that was the same color as the central square, MS was at chance level (52% correct; chance performance would be 50% correct).

There is a clear contrast between the absence of color experience for MS and the preservation of opponent cone mechanisms shown in his spectral sensitivity function (Fig. 8.1). This contrast is highlighted by the fact that MS was also able to detect chromatic boundaries, even when the colors on each side of the boundary were isoluminant (Heywood et al., 1991). Thus, he was able to distinguish whether or not a series of hues was in the correct chromatic order as long as the adjacent color patches abutted one another. If a small (5mm) gap was introduced between each color patch, his performance deteriorated to chance level. Heywood et al. (1991) accounted for this by suggesting that the salience of any chromatic border depends on the contrast between the hues on each side of the border; a random series of adjacent color patches have some very sharp borders (where very different hues abut each other) and some that are less sharp, whereas for an ordered series there should be no variation in the salience of each border. Comparable findings of preserved processing of certain aspects of wavelength without any phenomenal experience of seeing color have been reported for another patient, HJA, by Humphreys and his colleagues (Humphreys et al., 1992).

In cases of achromatopsia, then, there is evidence that some aspects of color processing mechanisms continue to function (Heywood et al., 1991; Humphreys et al., 1992). However, although some of the cortical mechanisms that show color sensitivity still operate, there is no subjective experience of color. This demonstrates that loss of phenomenal awareness can be selective to certain aspects of subjective experience; in this case, color.

PROSOPAGNOSIA

Prosopagnosic patients usually fail all tests of overt recognition of familiar faces (Hécaen & Angelergues, 1962; Meadows, 1974a). They cannot name the face, give the person's occupation or other biographical details, or even state whether or not a face is that of a familiar person (all faces seem unfamiliar). Even the most well-known faces may not be recognized, including famous people, friends, family, and the patient's own face when looking in a mirror. In contrast, recognition from nonfacial cues (such as voice, name, and sometimes even clothing or gait) is usually successful.

Although prosopagnosic patients no longer recognize familiar faces overtly, there is substantial evidence of covert recognition from physiological and behavioral measures (Bruyer, 1991; Young & de Haan, 1992). Bauer (1984) measured skin conductance whilst a prosopagnosic patient, LF, viewed a familiar face and listened to a list of five names. When the name belonged to the face LF was looking at, there was a greater skin conductance change than when someone else's name was read out. Yet if LF was asked to choose which name in the list was correct for the face, his performance was at chance level. These findings are summarized in Table 8.1, which makes it clear that the same effect was found to personally known faces (LF's family) and famous faces he would only have encountered in the mass media.

TABLE 8.1.
Percentages of famous names and family faces for which LF showed spontaneous naming, selection of the correct name from five alternatives, and maximum skin conductance response (GSR) to the correct name in the list of five alternatives (Bauer, 1984).

	Spontaneous naming	Name selection	Maximum GSR to correct name
Famous Faces:	0%	20%	60%
Family Faces:	0%	25%	63%

Bauer's (1984) study showed compellingly a difference between overt recognition that was at chance level for LF and some form of preserved covert recognition, as evidenced by his skin conductance responses. Comparable findings were reported by Tranel and Damasio (1985, 1988), using a different technique in which the patients simply looked at a series of familiar and unfamiliar faces. Skin conductance changes were greater to familiar than unfamiliar faces. This shows that it is possible to demonstrate a preserved electrophysiological response to the familiarity of the face alone that is a useful adjunct to Bauer's (1984) procedure.

A number of behavioral indices of covert recognition in prosopagnosia have also been developed. Eye movement scanpaths differ to familiar and unfamiliar faces, despite the absence of overt recognition (Rizzo, Hurtig, & Damasio, 1987). The patients are better at matching photographs of familiar faces than photographs of unfamiliar faces across transformations of orientation or age (de Haan, Young, & Newcombe, 1987a; Sergent & Poncet, 1990; Sergent & Signoret, 1992). When looking at a face, they are better at learning correct information than incorrect information about that person (Bruyer et al., 1983; de Haan et al., 1987a; de Haan, Young, & Newcombe, 1991; McNeil & Warrington,

1991; Sergent & Poncet, 1990; Sergent & Signoret, 1992; Young & de Haan, 1988). This superior learning of correct over incorrect information is found even for faces of people who have only been known since the patient's illness (de Haan et al., 1987a), and there is other evidence that also implies that despite the absence of overt recognition, representations of new faces can be created (Greve & Bauer, 1990; Tranel & Damasio, 1985, 1988).

Some of the most intriguing data come from priming and interference techniques. These are often used with normal subjects as tools for investigating relatively automatic aspects of recognition (Young & Bruce, 1991; Young & Ellis, 1989). Consider, for example, what happens if you are asked to classify a printed name as belonging to a politician or a nonpolitician and the name is presented together with an irrelevant face that you are asked to ignore. You find that you cannot successfully ignore the distractor face, and if it belongs to the wrong category, it will interfere with the speed of your response to the target name. This happens because recognition operates automatically; you cannot simply switch the recognition mechanism off by deciding not to recognize the distractor face. The same pattern of interference of distractor faces on the classification of name targets is found in some prosopagnosic patients, even though they do not recognize the distractor faces overtly (de Haan, Bauer, & Greve, 1992; de Haan et al., 1987a, 1987b; Sergent & Signoret, 1992).

In a typical priming experiment, subjects are asked to classify a target name as familiar or unfamiliar, but the target name is immediately preceded by a prime that may itself be a face or a name. On each trial in the experiment, then, a prime is presented and then a target name; subjects have to classify the target name as quickly as possible. Studies of this type have demonstrated that responses to the target name are facilitated if the prime is the same person as the target (e.g., if Ronald Reagan's face is used as the prime preceding the name Ronald Reagan) or a close associate (e.g., if Nancy Reagan's face is used as the prime preceding the name Ronald Reagan). This holds for prosopagnosic patients even though they do not recognize the face primes overtly (de Haan et al., 1992; Young, Hellawell, & de Haan, 1988). Moreover, it is possible to compare the size of the priming effect across face primes (which the patients mostly do not recognize overtly) and printed name primes (which they can recognize). These have been found to be exactly equivalent; the possibility of overt recognition of name primes makes no additional contribution to the priming effects observed.

Findings of covert recognition in prosopagnosia show responses based on the unique identities of familiar faces, even though overt recognition of these faces is not achieved. Prosopagnosia can thus be considered a selective deficit of access awareness in which awareness of recognition of familiar faces is lost.

AMNESIA

Following Warrington and Weiskrantz (1968, 1970), many studies have documented preserved priming effects in amnesia. Warrington and Weiskrantz (1968, 1970) asked amnesic patients to identify fragmented pictures or words and showed that subsequent identification of the same stimuli was facilitated. This type of finding can be obtained for amnesic patients even when they fail to remember having taken part in any previous testing sessions. The key point seems to be that the amnesics' memories are tested indirectly, in terms of facilitation of subsequent recognition of the same stimuli (Schacter, 1987). Direct tests, such as asking whether or not items were among those previously shown, lead to very poor performance. Hence, amnesics show a form of memory without awareness in which their performance can be affected by previous experiences they completely fail to remember overtly.

Amnesia can therefore be contrasted with prosopagnosia, as different forms of deficit affecting access awareness; in amnesia, there is impaired conscious access to memories of things that happened in the past.

ANOSOGNOSIA

Anosognosia is the term used by neurologists to refer to unawareness of an impairment caused by brain injury. Anosognosic patients may fail to comprehend their problems or even actively deny them. This can be considered to reflect defective monitoring.

The classic observations of unawareness of impairment were made by Von Monakow (1885) and Anton (1899), whose patients denied their own blindness or deafness. Although his name has become associated with denial of blindness, two of the three cases in Anton's report involved denial of deafness. The second case he described, however, was that of a 56-old woman who suffered a progressive visual field defect culminating in apparently complete blindness. She complained bitterly about her other problems but did not seem to take any notice of her blindness. When questioned about her vision, she maintained that it was not as good as when she was younger, but that she could still see.

There have been numerous subsequent reports of Anton's syndrome, as it has come to be known (McGlynn & Schacter, 1989). Most emphasize that the patients are unaware of their blindness, behave as if they can see, and may even confabulate visual experiences (Raney & Nielsen, 1942; Redlich & Dorsey, 1945). There are, though, some cases in which insight is achieved. Raney and Nielsen described a woman who, after a year of apparent lack of insight into her problems, exclaimed, "My God, I am blind! Just to think, I have lost my eyesight!"

In addition to blindness and deafness, a wide range of impairments can be subject to anosognosia. McGlynn and Schacter (1989) pointed out that patients who show unawareness of impairment are not necessarily demented and experience no global change of consciousness. Conversely, patients with multiple neuropsychological deficits, who are disoriented and confused, may nonetheless continue to achieve insight into their cognitive impairments (Parkin, Miller, & Vincent, 1987). Thus, there is evidence for a double dissociation between anosognosias and more general impairments of consciousness.

The view that unawareness of impairment cannot be attributed to global changes in consciousness is strengthened by the important observation that patients with more than one deficit may be unaware of one impairment yet perfectly well aware of others. Von Monakow (1885) noted that his patient complained of other problems, even though he was not aware of his visual impairment. Anton (1899) made similar observations and suggested that unawareness of impairment of a particular function is caused by a disorder at the highest levels of organization of *that* function.

This position has been further developed by Bisiach, Vallar, Perani, Papagno, and Berti (1986), who demonstrated dissociations between anosognosia for hemiplegia and anosognosia for hemianopia. Paralysis to one side of the body (hemiplegia) and blindness for half the field of vision (hemianopia) are relatively common consequences of brain injury. The fact that patients can show awareness of their hemiplegias but not their hemianopias, and vice versa, shows that anosognosia does not result from a general change in the patient's state of consciousness but can be specific to particular disabilities. Bisiach and his colleagues (1986, p. 480) concluded that "monitoring of the internal working is not secured in the nervous system by a general, superordinate organ, but is decentralized and apportioned to the different functional blocks to which it refers."

Unawareness of visual recognition impairments has also been noted. Bonhoeffer (1903) described a patient who showed no insight into his reading problems but was aware of his right hemianopia. Landis, Cummings, Christen, Bogen, and Imhof (1986) reported six cases of prosopagnosia, four of whom showed unconcern or denial of their face recognition problems. A detailed description of a case involving unawareness of impaired face recognition, SP, is given by Young, de Haan, and Newcombe (1990). SP showed a severe face-processing impairment that remained stable across a 20-month period of investigation, during which she had complete lack of insight into her face recognition difficulties. She was not distressed by her inability to recognize many familiar faces and maintained that she recognized faces "as well as before."

When directly confronted with her failure to recognize photographs of familiar faces, SP could only offer the suggestion that the photograph was a "poor likeness" or that she had "no recollection of having seen that person before." She had been a talented amateur artist, specializing in portraiture. After her illness, though, she could only recognize the portraits she had painted by careful deduction from the sitter's age, sex, background details, and so forth. When asked about her inability spontaneously to recognize her own paintings, SP commented only that she "had recognized them" and did not seem to think that there was anything unusual about the laborious method she used to achieve this. As Anton (1899) noted, anosognosic patients seem to lose their knowledge of what it was like to have the relevant ability; this is consistent with the problem being due to impaired monitoring.

In contrast to her lack of insight into her face recognition impairment, SP showed adequate insight into other physical and cognitive impairments produced by her illness, including poor memory, hemiplegia, and hemianopia. These findings bear out the point that unawareness of impairment does not result from any general change in the patient's state of consciousness but can be specific to particular disabilities; SP's lack of insight into her face recognition impairment involved a deficit-specific anosognosia.

Deficit-specific anosognosias may reflect impairment to monitoring mechanisms. We need to monitor our performance to correct errors we sometimes make, and also because different types of information must sometimes be intentionally combined and evaluated. For instance, if you see what looks to be your daughter going into the cinema when you thought she was at school, you must weigh up the degree of visual likeness of the seen person to your daughter against the probability that it could be her given your other knowledge. Studies of recognition errors made by normal people have found several examples of problems of this type (Young, 1993; Young, Hay, & Ellis, 1985). The point is neatly illustrated by Thomson (1986), who asked the daughter of an Australian couple to stand outside a London hotel when her parents thought she was in Australia. They recognized their daughter, but when she (deliberately) did not respond, her father apologized, "I am terribly sorry, I thought you were someone else."

Unawareness of impairment, then, does not result from any overall change in the patient's state of consciousness. Like other impairments of awareness, monitoring problems can be highly selective.

IMPLICATIONS AND COMPLICATIONS

The neuropsychological impairments discussed so far fit neatly against distinctions among phenomenal awareness, access awareness, and monitoring. Studies of blindsight show that the processing of visual stimuli can take place even when there is no phenomenal awareness of seeing them, and work on achromatopsia demonstrates that loss of phenomenal awareness can be selective even to certain aspects of subjective experience. Findings of covert recognition show that prosopagnosia can be considered a selective deficit of access awareness in which awareness of recognition of familiar faces is lost. In amnesia, there is impaired access to memories of things that happened in the past. Prosopagnosia and amnesia thus involve contrasting impairments of access awareness showing that, like phenomenal awareness, this can break down in different ways. Studies of anosognosia also show that, like the other impairments of awareness, monitoring problems can be highly selective. I have not discussed impairments of executive awareness as such, but problems in goal setting are often noted after frontal lobe lesions (Shallice, 1982). We know, then, that at least some of the problems caused by brain injury involve the disruption of different forms of awareness. In each case, there is no general perturbation of consciousness, but one aspect is lost, often in a highly selective way. Any adequate theory of consciousness needs to be able to accommodate this basic point.

These neuropsychological findings have important implications for our understanding of awareness. For example, they show the incorrectness of the assumption that awareness is integral to the operation of perceptual mechanisms. Instead, visual stimuli can be located even though they are not consciously perceived, boundaries defined only by differences in hue can be detected without the experience of seeing color, the identities of faces that are not consciously recognized can influence responses to peoples' names, and so on.

Although this is a step forward, it does not explain how awareness is achieved by neural mechanisms. Three main alternative conceptions remain plausible: (a) Awareness only occurs when some form of threshold is crossed, (b) awareness is dependent on a different neural system or systems to those used for perceptual processing, or (c) awareness is only associated with certain types of perceptual processing. These are not mutually exclusive possibilities; any or all might be correct.

Clues can be found in observations that show that these forms of loss of awareness after brain injury are not always absolute. In cases of blindsight, there are differences in the degree to which all forms of visual experience are lost in the blind area. Soon after surgery, DB was not aware of stimuli in his blind field at all, except that in some regions moving stimuli could produce peculiar radiating lines. Later, he said that he "knew something was there" and roughly where it was, but that he did not in any sense see it (Weiskrantz, 1980). Another patient, G, perceived some light flashes presented in the blind area as dark shadows (a very abnormal response), but more intense stimulation could give the sensation of a well-localized bright flash (Barbur, Ruddock, & Waterfield, 1980). There are even reports that practice can lead to the restitution of visual experience in some cases (Zihl, 1981; Zihl & von Cramon, 1979).

Some of the most tantalizing findings were noted by Torjussen (1976, 1978), who found that patients with right hemianopias reported seeing a circle when it was presented so as to fall half in the blind and half in the sighted region, yet they reported seeing a semicircle if this was presented in the sighted region and nothing if a semicircle was presented in the blind right visual field.

Torjussen's findings are consistent with other examples in which there are interactions between stimuli presented in sighted and blind parts of the visual field (Marzi et al., 1986; Pizzamiglio et al., 1984; Weiskrantz, 1990), but they are particularly significant because they show that the interaction between the stimuli in the blind and sighted parts of the visual field can affect what people report that they see in the blind region. Notice that presenting a semicircle in the blind field alone was insufficient for it to be perceived, yet essentially the same stimulus was reported as having been seen when it formed a part of a complete figure that extended into the sighted field.

What may be a related example comes from recent work on prosopagnosia. Sergent and Poncet (1990) observed that their patient, PV, could achieve overt recognition of some faces if several members of the same semantic category were presented together. This only happened when PV could determine the category herself. For the categories PV could not determine, she continued to fail to recognize the faces overtly even when the occupational category was pointed out to her. This phenomenon of overt recognition provoked by multiple exemplars of a semantic category has been replicated with case PH by de Haan and his colleagues (1991) and with case PC by Sergent and Signoret (1992). PV, PH, and PC were all noted to be very surprised at being able to recognize faces overtly.

Sergent and Poncet (1990, p. 1000) suggested that their demonstration shows that "neither the facial representations nor the semantic information were critically disturbed in PV, and her prosopagnosia may thus reflect faulty connections between faces and their memories." They hypothesized that the simultaneous presentation of several members of the same category may have temporarily raised the activation level above the appropriate threshold.

Such findings in blindsight and prosopagnosia show that the boundary between awareness and lack of awareness is not as completely impassable as it seems to the patients' everyday experience. The fact that certain types of stimulation can trigger experiences they no longer enjoy routinely fits readily with a model of the type preferred by Sergent and Poncet (1990), in which activation must cross some form of threshold before it can result in awareness. However, it must be emphasized that the circumstances under which this has been found to happen are at present very limited.

There are other impairments that show that the conscious or nonconscious boundary is not always clearly marked. Unilateral neglect is a good example; in some ways it seems more like a distortion than a loss of awareness. Patients with unilateral neglect fail to respond to stimuli located contralaterally to the side of their brain lesion. The condition is more common after lesions of the right cerebral hemisphere, when left-sided stimuli are neglected (Heilman, 1979).

Neglect has often been discussed as if the fact that patients do not respond to left-sided stimuli implies that they are not aware of them and has been claimed to form a parallel phenomenon to blindsight. In an ingenious experiment, Marshall and Halligan (1988) showed a drawing of a house in which bright red flames were emerging from the left-hand side to PS, a patient with left-sided neglect. Although she correctly identified the drawing as a house, PS failed to report that it was on fire and asserted that this drawing was the same as another drawing that had no flames. Yet when asked which house she would prefer to live in, PS consistently tended to choose the nonburning picture, even though she considered the questions silly "because they're the same." The information that the house was on fire was apparently registered at a nonconscious level.

In later trials, PS was shown a drawing in which flames emerged from the right-hand side of the house that she immediately commented upon. Subsequently, she finally noticed the fire in the house with the left-sided flames, exclaiming, "Oh my God, this one's on fire!" Thus, the failure of conscious recognition of the left-sided flames could be overcome by appropriate cueing.

A number of similar findings of processing without awareness for neglected stimuli have been reported recently (Berti & Rizzolatti, 1992; McGlinchey-Berroth, Milberg, Verfaellie, Alexander, & Kilduff, 1993). However, other studies have described phenomena that highlight the complex and paradoxical nature of the breakdown of perceptual awareness in neglect, suggesting the need for caution in the interpretation of this disorder (Bisiach & Rusconi, 1990; Young, Hellawell, & Welch, 1992). Bisiach and Rusconi found that neglect of features on the left sides of objects sometimes persisted when their patients had traced around the left sides with their fingers, and other reports indicate that neglect of the left side of a stimulus can occur even when it must have fallen in the patient's perimetrically intact right visual field (Bisiach, Luzzatti, & Perani, 1979; Ellis, Flude, & Young, 1987; Kinsbourne & Warrington, 1962; Young, de Haan, & Newcombe, 1990; Young, Newcombe, & Ellis, 1991).

Further examples come from our study of BQ, who had a severe and longstanding left-sided neglect (Young et al., 1992). When asked to recognize crudely constructed chimaeric stimuli, BQ was very poor at identifying the left half of each chimaeric, yet she could recognize these same left sides if they were presented in isolation (see Fig. 8.2 and Table 8.2). Even when shown how the chimaerics were constructed and told to report both sides, BQ would often deny that she was looking at anything other than a single object or face and identify only the right side.

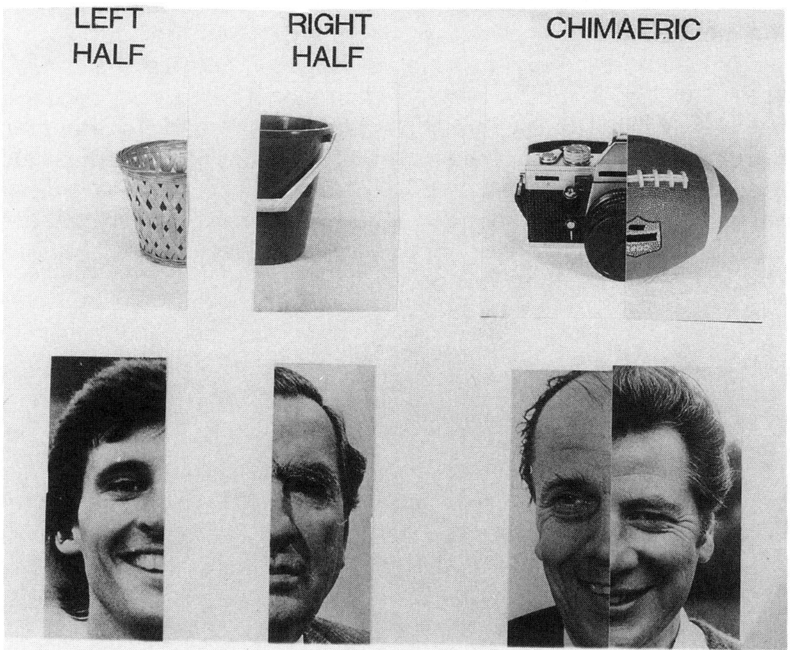

FIG. 8.2. Examples of left and right half-objects and half-faces, chimaeric objects, and chimaeric faces. Reproduced by permission of Oxford University Press from Young, A. W., Hellawell, D. J., & Welch, J. (1992). Neglect and visual recognition. *Brain, 115,* 51–71.

TABLE 8.2.
Error Rates for BQ's Recognition of Left and Right Half-Objects and Half-Faces and the Left and Right Sides of Chimaeric Objects and Faces. Reproduced by permission of Oxford University Press from Young, A. W., Hellawell, D. J., & Welch, J. (1992). Neglect and visual recognition. *Brain, 115,* 51–71.

	Half-stimuli		Chimaeric Stimuli	
	Left half	Right half	Left side	Right side
Objects:	1/20	0/20	14/20	2/20
Faces:	2/20	0/20	20/20	0/20

Eventually, BQ became exasperated with our persistence in maintaining that the stimuli were chimaerics and tried to convince us that there was only a single object. She identified the chimaeric shown in Fig. 8.3 as "a bowl." When asked what was on the left, she insisted that the whole object was a bowl, and to demonstrate this she spontaneously traced around the rim with her finger along the trajectory shown in Fig. 8.3. To do this, she must have needed some visual guidance for the finger movement, yet she made no comment even when her finger passed across the left side of the chimaeric, where there was no trace of the rim.

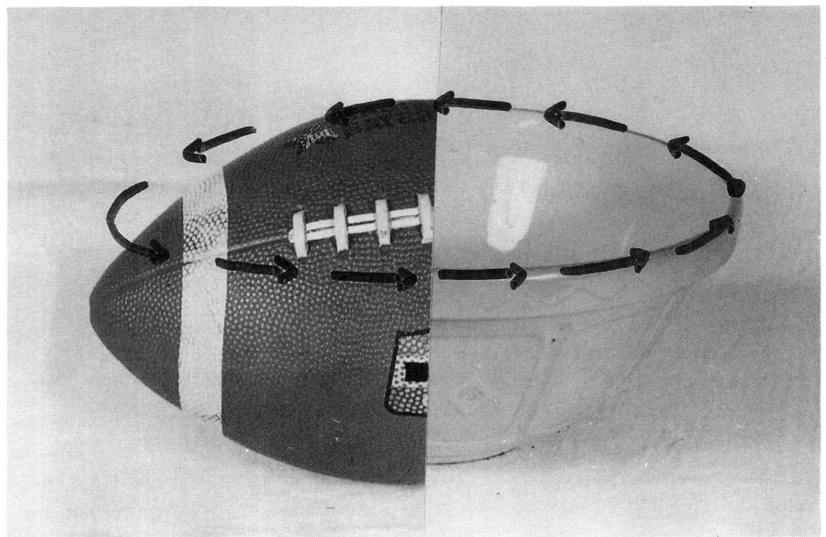

FIG. 8.3. BQ's demonstration that only a single object was present, by tracing the bowl's rim. Reproduced by permission of Oxford University Press from Young, A. W., Hellawell, D. J., & Welch, J. (1992). Neglect and visual recognition. *Brain, 115,* 51–71.

When she was asked to trace around the contours of other chimaeric objects, BQ would sometimes trace nonexistent contours on the object's left (as in Fig. 8.3), but more often her tracing was reasonably accurate. Yet when asked to identify the parts of chimaeric objects whose contours she had traced correctly, BQ still relied on the right side of each.

A similar phenomenon was found with chimaeric faces. If we pointed to parts of the chimaeric on the left, BQ could describe them accurately, yet she nearly always identified the chimaeric as the right half-face. With a chimaeric in which the left half came from Mick Jagger's face and the right half from Roger Moore, she described accurately the shape of the left eye, the long hair over the left ear, the full lips on the left, and the stubble on the left chin when we pointed to these features; yet even after describing all of these features of Mick Jagger's face successfully, BQ identified the chimaeric only as Roger Moore, and did not accept that part of another person's face was present. On another occasion she showed partial insight; when made to describe parts of a face chimaeric with the left half of Michael Parkinson and the right half of Terry Wogan, she commented, "It's Terry Wogan . . . but there's a touch of a Picasso about him." There was only occasional full insight into the fact that the stimuli were chimaeric; for example, with a face chimaeric of the left half of Elvis Presley and the right half of Steve Davis, BQ commented that it was "Elvis Presley . . . but there's a bit of Steve Davis somewhere", and then noticed there were "two faces stuck together."

These observations show that in cases of unilateral neglect, it is not at all easy to distinguish what is seen from what is not seen; even stimuli that have been accurately described by the patients in one task (and of which they were therefore presumably aware) will not always be reported or seem directly to influence performance in another task.

PERCEPTION AND ACTION

Neuropsychological data are also important to understanding the functions of awareness. In fact, they fit the common sense idea that awareness supports intentional action. For example, despite the extensive range of covert effects demonstrated in the laboratory, prosopagnosic patients do not act as if they recognize faces in everyday life. This is not a trivial point; one could imagine that a prosopagnosic patient with covert recognition might find himself greeting people in the street without knowing why. This has not been found. Similarly, patients who show blindsight in laboratory tasks do not respond to objects located in their blind field in everyday life; in the laboratory they must be instructed to guess.

What, then, is the functional significance of covert effects? It seems that they reflect aspects of visual information processing that can proceed without conscious experience. Why?

Part of the answer lies in the delicate balance between speed of response and flexibility of response. The purpose of perceptual systems is to create representations of external events that can permit effective action in the world that an organism inhabits. But flexibility of response requires more sophisticated representations of events that take longer to compute. Consider the favorite textbook comparison to the frog, whose retinal outputs include bug detectors that allow it to make very fast responses to flies. Our own visual system is much less specialized at the retinal level, allowing increased flexibility of response at the cost of loss of speed. Some types of covert effect can thus be seen to reflect automatic components built in to the visual system to reduce some of this loss of speed, for example, by predicting what is likely to follow and setting up preparatory responses.

The costs of flexibility become especially marked when time is spent not only in constructing an adequate representation but also in weighing up possible alternative actions. An obvious way to balance the competing demands of flexibility and speed is to allow most actions to run off under automatic control and to involve mechanisms that can allow greater choice only when these are needed. This is a particularly convenient solution for the nervous system in which many actions (like breathing) can be safely left under "automatic pilot" much of the time and only require occasional conscious intervention (if you are about to stick your head underwater, and so on). It is doubly convenient because nervous systems must evolve, and one way to do this is to add extra levels of control to mechanisms that are already sufficient for many purposes. Such points have been recognized for many years both in neuropsychology (Hughlings Jackson, 1884; Luria, 1973) and in experimental psychology (Baars, 1988; Shiffrin & Schneider, 1977). They come as no surprise to nonpsychologists, who often note that the performance of skilled actions is disrupted rather than facilitated if one attends to them too carefully.

A compelling demonstration of the force of this general position comes from the dissociation of action and conscious perception found in the elegant work of Milner, Goodale, and their colleagues (Goodale & Milner, 1992; Goodale, Milner, Jakobson, & Carey, 1991; Milner et al., 1991). They have made a very thorough investigation of a case of visual form agnosia, a neuropsychological condition in which a severe visual recognition impairment is found in the context of defective processing of basic properties involved in shape perception (Benson & Greenberg, 1969; Efron, 1968; Sparr, Jay, Drislane, & Venna, 1991).

Milner and his colleagues' (1991) patient, DF, was severely impaired when she was asked to judge the orientation of a slot by manually rotating a visible matching slot to the same inclination. In the top row of Fig. 8.4, each line represents one of her attempts to do this, in terms of the final orientation she chose for the matching slot; the mean error is some 40°. Yet when DF was asked to put her hand into the slot, she immediately positioned it correctly, as can be seen from the bottom row of Fig. 8.4, where the mean error is less than 5°. Other studies showed that DF could rotate a slot into a verbally specified position as accurately as normal subjects if she had her eyes closed and that she shaped her fingers appropriately for the size of an object she was about to pick up, even though her ability to make overt size judgments was poor (Goodale & Milner, 1992; Goodale et al., 1991).

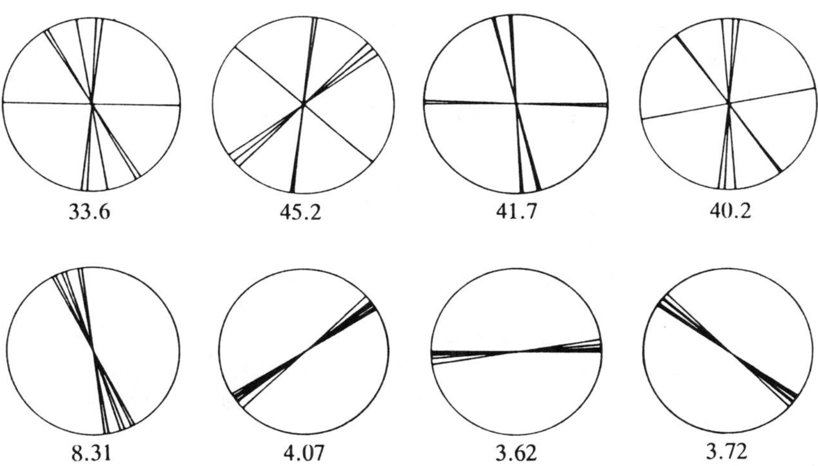

FIG. 8.4. Accuracy of orientation judgments made manually by rotating a visible matching slot (top row) against terminal hand orientation in reaching to an equivalently oriented (bottom row) for patient DF. The four target orientations (columns, from left to right) are vertical, 45°, horizontal, and 135°. Reproduced by permission of Oxford University Press from Milner, A. D., Perrett, D. I., Johnston, R. S., Benson, P. J., Jordan, T. R., Heeley, D. W., Bettucci, D., Mortara, F., Mutani, R., Terazzi, E., & Davidson, D. L. W. (1991). Perception and action in 'visual form agnosia.' *Brain, 114,* 405–428.

Goodale and Milner (1992) pointed out that cases like DF, who can make accurate movements directed at objects she does not seem to be seeing at all well, contrast with other cases involving bilateral parietal damage in which actions on the world are very defective even though shape processing and visual recognition are intact. They propose that such differences are "largely a reflection of the specific transformations of input required by perception and action", and that functional modularity in vision "extends from input right through to output" (Goodale & Milner, 1992, p. 24).

An important point about Milner and Goodale's work is that they have demonstrated that it is only certain types of action that are preserved for DF. In Fig. 8.4, for example, all of the responses are made manually; the difference is that the accurate responses involve a well-practiced everyday movement that can be run off without conscious control (putting your hand into something), whereas the inaccurate responses arise in a task that needs continual conscious intervention (judging whether or not two orientations are the same and how to reduce any difference between them).

This point is well made by Frith (1987), who has drawn a distinction between two routes to action involving a *stimulus intention* (a purposive action that the subject believes is required by the stimulus) and a *willed intention* (an action determined by an overall plan or goal). Frith argued that although many actions involve a combination of stimulus and willed intentions, the nervous system respects the underlying difference.

UNITY OR DISUNITY OF CONSCIOUS EXPERIENCE

We have seen that neuropsychological findings show that awareness is not integral to certain types of information processing and that part of the reason for this is that conscious mechanisms are not needed for many routine purposes. Can we say anything more about the level at which conscious mechanisms operate?

William James (1890, p. 144) pointed out that "the *distribution* of consciousness shows it to be exactly such as we might expect in an organ added for the sake of steering a nervous system grown too complex to regulate itself." It is therefore tempting to equate consciousness with the functioning of the executive systems involved in higher-order control, but this is probably a mistake. Social psychologists have demonstrated that there are many circumstances in which we have little insight into the true reasons for our actions (Nisbett & Wilson, 1977); a point that was often made in 19th century psychology and psychiatry (Ellenberger, 1970). Instead, it is more useful to see the contents of consciousness as "an attempt to make sense of as much data as possible at the most functionally useful level" (Marcel, 1983, p. 238). A particularly appealing hypothesis is that the contents of consciousness correspond to the information that is made available to executive mechanisms (Morris & Hampson, 1983). This could include the feedback of information that these executive mechanisms need to continue to make available to themselves (i.e., executive awareness) as well as information from perceptual, memory, and monitoring systems (phenomenal awareness, access awareness, and monitoring). From this perspective, the different types of loss of awareness caused by brain injuries are much less of a surprise (Morris, 1992).

A related issue concerns whether there are single or multiple conscious mechanisms. Many discussions of conscious experience begin by emphasizing its subjective unity and continuity (James, 1890), and some have argued that materialist theories cannot account for this unity (Tallis, 1991). Similarly, although Morris (Morris, 1992; Morris & Hampson, 1983) maintained that the contents of consciousness correspond to the information that is made available to executive mechanisms, he seemed to think that there is only a single executive system. Likewise, an influential review by Schacter and his colleagues (1988) argued that covert effects can be considered to reflect disconnection of specialized processing modules from a single conscious mechanism.

I began by being very attracted to such views, but the evidence from neuropsychology now seems to me to imply that we should be suspicious of theories that treat consciousness as if it were a single, homogeneous entity. The list of different dissociations that can be observed is increasing rapidly; if these were all due to damaged interaction with a unitary conscious mechanism, it is becoming hard to see how our brains could contain all the wiring needed for its inputs and outputs. There are no neurological grounds for supposing that there is any central bottleneck in the nervous system.

Similarly, if there were a unitary conscious mechanism, why should different aspects of awareness (such as phenomenal awareness, access awareness, or monitoring) be differentially affected in different cases? We might have expected that the existence of a unitary conscious mechanism would predict that all impairments would simultaneously compromise phenomenal awareness, access awareness and monitoring to equivalent extents. Instead, one has to wonder whether the subjective unity of conscious phenomena is not largely illusory. Even if there were multiple conscious mechanisms, why should they not "feel" unified in operation? What else could they feel like?

UNDERSTANDING COVERT EFFECTS

An important issue concerns the value of explicit models. To illustrate this, I return to the example of covert recognition in prosopagnosia. Some of the functional mechanisms underlying face recognition have been carefully investigated in studies of normal subjects (Bruce, 1988; Young & Bruce, 1991), and their properties have been neatly simulated in a computer model developed by Burton, Bruce, and Johnston (1990). This computer simulation mimics the patterns of priming and other effects found in normal people. The point of interest is here that Burton and his colleagues' (1990) computer simulation of the normal system shows preserved associative priming without explicit classification of face inputs if the connection strengths between two of its pools of units (FRUs and PINs) are halved (Burton, Young, Bruce, Johnston, & Ellis, 1991). This makes the finding of this pattern in some cases of prosopagnosia much less mysterious. The problem of understanding how covert responses can be preserved when there is no overt discrimination may thus be less intractable than it at first appeared, but of course we have not provided any solution to the more philosophical problems concerning awareness (we do not claim that the computer is aware just because it passes an arbitrary threshold).

This leads to the issue of damage and disconnection accounts of covert effects. A distinction is often made in neurology between effects that are due to damage to a functional system and effects due to disconnection of one system from another (Geschwind, 1965a, 1965b). The same distinction has been applied to work on covert processing after brain injury, where patterns of preserved abilities have been taken to reflect the disconnection of intact processing systems from mechanisms that can sustain awareness, rather than damage to processing systems per se (de Haan et al., 1987a; Schacter et al., 1988; Young & de Haan, 1988). Our simulation (Burton et al., 1991) shows the inadequacy of this contrast. The same simulation can equally well be taken to model disconnection (it is connection strengths that are changed to impair the model's functions) or to model damage (changing one set of connection strengths affects all components because of their interconnectivity and therefore damages the functioning of the system considered as a whole); it is simply a question of how one prefers to describe it.

MULTIPLE FUNCTIONS OF CONSCIOUSNESS

Finally, I want to make another obvious point, that consciousness has multiple functions. I have concentrated here on cognitive functions of consciousness in perception and the control of action. But consciousness is just as involved in creating a sense of self, and it has marked emotional components, too.

Some idea of the importance of this can be found in the effects of brain injuries that result in bizarre delusions. My colleagues and I have recently had the opportunity to work with a number of people who have experienced delusional beliefs. We have been particularly intrigued by the Capgras delusion in which patients claim that people (usually close relatives) have been replaced by identical or near-identical impostors (Capgras & Reboul-Lachaux, 1923) and the Cotard delusion in which patients claim that they are dead (Cotard, 1882). The fact that these delusional beliefs can follow brain injury (Cutting, 1991; Lewis, 1987; Young, Reid, Wright, & Hellawell, 1993; Young, Robertson, Hellawell, de Pauw, & Pentland, 1992) renders psychodynamic accounts unsatisfying as a full explanation, and we have sought to tease out the roles of neuropsychological factors.

We think that the Capgras and Cotard delusions each reflect an interaction of impairments at two levels. One set of contributory factors involve anomalous perceptual experience the other factors lead to an incorrect interpretation of this.

Lewis (1987) and Ellis and Young (1990) suggested that the Capgras delusion involves damage to neuroanatomical pathways responsible for appropriate emotional reactions to familiar visual stimuli. On this account, the Capgras delusion typically involves close relatives because these would normally produce the strongest reactions and hence suffer the greatest discrepancy. A similar process might well be involved in the Cotard delusion, where patients commonly report that they must be dead because they "feel nothing inside." The delusions would then represent the patients' attempts to make sense of these puzzling changes (Wright, Young, & Hellawell, 1993; Young et al., 1993).

How might this happen? Kaney and Bentall (1989) and Candido and Romney (1990) showed that people with persecutory delusions tend to attribute negative events to external rather than internal causes, whereas depressed people tend to attribute them to internal causes. The relevance of these findings is that it is quite common for the Cotard delusion to arise in the setting of a depressive illness and for the Capgras delusion to be accompanied by persecutory delusions and suspiciousness (Enoch & Trethowan, 1991). Hence, the persecutory delusions and suspiciousness that are often noted in Capgras cases could contribute to the patients mistaking a change in themselves for a change in others ("they are impostors"), whereas people who are depressed might exaggerate the negative effects of a similar change while correctly attributing it to themselves ("I am dead").

The purpose of this digression to the outer limits of psychiatry is to make clear that there are many more observations that need to be incorporated before we have any detailed understanding of the neuropsychology of awareness. In many ways, our knowledge of brain injuries that can affect awareness is at present grossly inadequate. But what we do know suggests that the neuropsychological approach has a lot to offer in understanding consciousness and awareness.

ACKNOWLEDGMENTS

I gratefully acknowledge the support provided by ESRC grant R000231922. Thanks to Dr. Charlie Heywood and Oxford University Press for permission to reproduce Figure 8.1, to Oxford University Press for Figure 8.2, Table 8.2, and Figure 8.3 and to Professor David Milner and Oxford University Press for Figure 8.4.

REFERENCES

Allport, A. (1988). What concept of consciousness? In A. J. Marcel & E. Bisiach (Eds.), *Consciousness in contemporary science* (pp. 159–182). Oxford: Oxford University Press.

Anton, G. (1899). Über die Selbstwahrnemung der Herderkrankungen des Gehirns durch den Kranken bei Rindenblindheit und Rindentaubheit. *Archiv für Psychiatrie und Nervenkrankheiten, 32,* 86–127.

Baars, B. J. (1988). *A cognitive theory of consciousness.* Cambridge: Cambridge University Press.

Barbur, J. L., Ruddock, K. H., & Waterfield, V. A. (1980). Human visual responses in the absence of the geniculo-calcarine projection. *Brain, 103,* 905–928.

Bauer, R. M. (1984). Autonomic recognition of names and faces in prosopagnosia: A neuropsychological application of the guilty knowledge test. *Neuropsychologia, 22,* 457–469.

Benson, D. F., & Greenberg, J. P. (1969). Visual form agnosia: A specific defect in visual discrimination. *Archives of Neurology, 20,* 82–89.

Berti, A., & Rizzolatti, G. (1992). Visual processing without awareness: Evidence from unilateral neglect. *Journal of Cognitive Neuroscience, 4,* 345–351.

Bisiach, E., Luzzatti, C., & Perani, D. (1979). Unilateral neglect, representational schema and consciousness. *Brain, 102,* 608–618.

Bisiach, E., & Rusconi, M. L. (1990). Break-down of perceptual awareness in unilateral neglect. *Cortex, 26,* 643–649.

Bisiach, E., Vallar, G., Perani, D., Papagno, C., & Berti, A. (1986). Unawareness of disease following lesions of the right hemisphere: Anosognosia for hemiplegia and anosognosia for hemianopia. *Neuropsychologia, 24,* 471–482.

Block, N. (1991). Evidence against epiphenomenalism. *Behavioral and Brain Sciences, 14,* 670–672.

Bonhoeffer, K. (1903). Casuistische Beiträge zur Aphasialehre. *Archiv für Psychiatrie und Nervenkrankheiten, 37,* 564–597.

Bruce, V. (1988). *Recognising faces.* London: Lawrence Erlbaum Associates.

Bruyer, R. (1991). Covert face recognition in prosopagnosia: A review. *Brain and Cognition, 15,* 223–235.

Bruyer, R., Laterre, C., Seron, X., Feyereisen, P., Strypstein, E., Pierrard, E., & Rectem, D. (1983). A case of prosopagnosia with some preserved covert remembrance of familiar faces. *Brain and Cognition, 2,* 257–284.

Burton, A. M., Bruce, V., & Johnston, R. A. (1990). Understanding face recognition with an interactive activation model. *British Journal of Psychology, 81,* 361–380.

Burton, A. M., Young, A. W., Bruce, V., Johnston, R., & Ellis, A. W. (1991). Understanding covert recognition. *Cognition, 39,* 129–166.

Candido, C. L., & Romney, D. M. (1990). Attributional style in paranoid vs. depressed patients. *British Journal of Medical Psychology, 63,* 355–363.

Capgras, J., & Reboul-Lachaux, J. (1923). L'illusion des "sosies" dans un délire systématisé chronique. *Bulletin de la Société Clinique de Médicine Mentale, 11,* 6–16.

Cotard, J. (1882). Du délire des négations. *Archives de Neurologie, 4,* 150–170, 282–295.

Cowey, A., & Stoerig, P. (1991). The neurobiology of blindsight. *Trends in Neurosciences, 14,* 140–145.

Cowey, A., & Stoerig, P. (1992). Reflections on blindsight. In A. D. Milner & M. D. Rugg (Eds.), *The neuropsychology of consciousness* (pp. 11–37). London: Academic Press.

Cutting, J. (1991). Delusional misidentification and the role of the right hemisphere in the appreciation of identity. *British Journal of Psychiatry, 159,* 70–75.

de Haan, E. H. F., Bauer, R. M., & Greve, K. W. (1992). Behavioural and physiological evidence for covert face recognition in a prosopagnosic patient. *Cortex, 28,* 77–95.

de Haan, E. H. F., Young, A., & Newcombe, F. (1987a). Face recognition without awareness. *Cognitive Neuropsychology, 4,* 385–415.

de Haan, E. H. F., Young, A., & Newcombe, F. (1987b). Faces interfere with name classification in a prosopagnosic patient. *Cortex, 23,* 309–316.

de Haan, E. H. F., Young, A. W., & Newcombe, F. (1991). Covert and overt recognition in prosopagnosia. *Brain, 114,* 2575–2591.

Efron, R. (1968). What is perception? In R. S. Cohen & M. W. Wartofsky (Eds.), *Boston studies in the philosophy of science, 4* (pp. 137–173). Dordrecht: Reidel.

Ellenberger, H. F. (1970). *The discovery of the unconscious: The history and evolution of dynamic psychiatry.* London: Allen Lane The Penguin Press.

Ellis, A. W., Flude, B. M., & Young, A. W. (1987). "Neglect dyslexia" and the early visual processing of letters in words and nonwords. *Cognitive Neuropsychology, 4,* 439–464.

Ellis, H. D., & Young, A. W. (1990). Accounting for delusional misidentifications. *British Journal of Psychiatry, 157,* 239–248.

Enoch, M. D., & Trethowan, W. H. (1991). *Uncommon psychiatric syndromes* (3rd ed.). Oxford: Butterworth-Heinemann.

Frith, C. D. (1987). The positive and negative symptoms of schizophrenia reflect impairments in the perception and initiation of action. *Psychological Medicine, 17,* 631–648.

Geschwind, N. (1965a). Disconnexion syndromes in animals and man. Part I. *Brain, 88,* 237–294.

Geschwind, N. (1965b). Disconnexion syndromes in animals and man. Part II. *Brain, 88,* 585–644.

Goodale, M. A., & Milner, A. D. (1992). Separate visual pathways for perception and action. *Trends in Neurosciences, 15,* 20–25.

Goodale, M. A., Milner, A. D., Jakobson, L. S., & Carey, D. P. (1991). A neurological dissociation between perceiving objects and grasping them. *Nature, 349,* 154–156.

Greve, K. W., & Bauer, R. M. (1990). Implicit learning of new faces in prosopagnosia: An application of the mere-exposure paradigm. *Neuropsychologia, 28,* 1035–1041.

Hécaen, H., & Angelergues, R. (1962). Agnosia for faces (prosopagnosia). *Archives of Neurology, 7,* 92–100.

Heilman, K. M. (1979). Neglect and related disorders. In K. M. Heilman & E. Valenstein (Eds.), *Clinical neuropsychology* (pp. 268–307). New York: Oxford University Press.

Heywood, C. A., Cowey, A., & Newcombe, F. (1991). Chromatic discrimination in a cortically blind observer. *European Journal of Neuroscience, 3,* 802–812.

Hughlings Jackson, J. (1884). Evolution and dissolution of the nervous system: Croonian lectures delivered at the Royal College of Physicians, March 1884. In J. Taylor (Ed.), *Selected writings of John Hughlings Jackson, Vol. 2* (1958, pp. 45–75). London: Staples Press.

Humphreys, G. W., Troscianko, T., Riddoch, M. J., Boucart, M., Donnelly, N., & Harding, G. F. A. (1992). Covert processing in different visual recognition systems. In A. D. Milner & M. D. Rugg (Eds.), *The neuropsychology of consciousness* (pp. 39–68). London: Academic Press.

James, W. (1890). *The Principles of psychology, Vol. 1.* New York: Dover.

Kaney, S., & Bentall, R. P. (1989). Persecutory delusions and attributional style. *British Journal of Medical Psychology, 62,* 191–198.

Kinsbourne, M., & Warrington, E. K. (1962). A variety of reading disability associated with right hemisphere lesions. *Journal of Neurology, Neurosurgery, and Psychiatry, 25,* 339–344.

Landis, T., Cummings, J. L., Christen, L., Bogen, J. E., & Imhof, H.-G. (1986). Are unilateral right posterior cerebral lesions sufficient to cause prosopagnosia? Clinical and radiological findings in six additional patients. *Cortex, 22,* 243–252.

Lewis, S. W. (1987). Brain imaging in a case of Capgras' Syndrome. *British Journal of Psychiatry, 150,* 117–121.

Luria, A. R. (1973). *The working brain: An introduction to neuropsychology* (B. Haigh, Trans.). Harmondsworth: Allen Lane The Penguin Press.

Marcel, A. J. (1983). Conscious and unconscious perception: An approach to the relations between phenomenal experience and perceptual processes. *Cognitive Psychology, 15,* 238–300.

Marshall, J. C., & Halligan, P. W. (1988). Blindsight and insight in visuo-spatial neglect. *Nature, 336,* 766–767.

Marzi, C. A., Tassinari, G., Agliotti, S., & Lutzemberger, L. (1986). Spatial summation across the vertical meridian in hemianopics: A test of blindsight. *Neuropsychologia, 24,* 749–758.

McGlinchey-Berroth, R., Milberg, W. P., Verfaellie, M., Alexander, M., & Kilduff, P. T. (1993). Semantic processing in the neglected visual field: Evidence from a lexical decision task. *Cognitive Neuropsychology, 10,* 79–108.

McGlynn, S., & Schacter, D. L. (1989). Unawareness of deficits in neuropsychological syndromes. *Journal of Clinical and Experimental Neuropsychology, 11,* 143–205.

McNeil, J. E., & Warrington, E. K. (1991). Prosopagnosia: A reclassification. *Quarterly Journal of Experimental Psychology, 43A,* 267–287.

Meadows, J. C. (1974a). The anatomical basis of prosopagnosia. *Journal of Neurology, Neurosurgery, and Psychiatry, 37,* 489–501.

Meadows, J. C. (1974b). Disturbed perception of colours associated with localized cerebral lesions. *Brain, 97,* 615–632.

Milner, A. D., Perrett, D. I., Johnston, R. S., Benson, P. J., Jordan, T. R., Heeley, D. W., Bettucci, D., Mortara, F., Mutani, R., Terazzi, E., & Davidson, D. L. W. (1991). Perception and action in "visual form agnosia". *Brain, 114,* 405–428.

Mollon, J. D., Newcombe, F., Polden, P. G., & Ratcliff, G. (1980). On the presence of three cone mechanisms in a case of total achromatopsia. In G. Verriest (Ed.), *Colour vision deficiencies, V* (pp. 130–135). Bristol: Hilger.

Morris, P. E. (1992). Cognition and consciousness. *The Psychologist: Bulletin of the British Psychological Society, 5,* 3–8.

Morris, P. E., & Hampson, P. J. (1983). *Imagery and consciousness.* London: Academic Press.

Newcombe, F., & Ratcliff, G. (1974). Agnosia: A disorder of object recognition. In F. Michel & B. Schott (Eds.), *Les syndromes de disconnexion calleuse chez l'homme* (pp. 317–341). Lyon: Colloque Internationale de Lyon.

Nisbett, R. E., & Wilson, T. D. (1977). Telling more than we can know: Verbal reports on mental processes. *Psychological Review, 84,* 231–259.

Parkin, A. J., Miller, J., & Vincent, R. (1987). Multiple neuropsychological deficits due to anoxic encephalopathy: A case study. *Cortex, 23,* 655–665.

Pizzamiglio, L., Antonucci, G., & Francia, A. (1984). Response of the cortically blind hemifields to a moving visual scene. *Cortex, 20,* 89–99.

Pöppel, E., Held, R., & Frost, D. (1973). Residual visual function after brain wounds involving the central visual pathways in man. *Nature, 243,* 295–296.

Raney, A. A., & Nielsen, J. M. (1942). Denial of blindness (Anton's Symptom). *Bulletin of Los Angeles Neurological Society, 7,* 150–151.

Ratcliff, G., & Newcombe, F. (1982). Object recognition: Some deductions from the clinical evidence. In A. W. Ellis (Ed.), *Normality and pathology in cognitive functions* (pp. 147–171). London: Academic Press.

Redlich, F. C., & Dorsey, J. F. (1945). Denial of blindness by patients with cerebral disease. *Archives of Neurology and Psychiatry, 53,* 407–417.

Rizzo, M., Hurtig, R., & Damasio, A. R. (1987). The role of scanpaths in facial recognition and learning. *Annals of Neurology, 22,* 41–45.

Schacter, D. L. (1987). Implicit memory: History and current status. *Journal of Experimental Psychology: Learning, Memory, and Cognition, 13,* 501–518.

Schacter, D. L., McAndrews, M. P., & Moscovitch, M. (1988). Access to consciousness: Dissociations between implicit and explicit knowledge in neuropsychological syndromes. In L. Weiskrantz (Ed.), *Thought without language* (pp. 242–278). Oxford: Oxford University Press.

Sergent, J., & Poncet, M. (1990). From covert to overt recognition of faces in a prosopagnosic patient. *Brain, 113,* 989–1004.

Sergent, J., & Signoret, J.-L. (1992). Implicit access to knowledge derived from unrecognized faces in prosopagnosia. *Cerebral Cortex, 2,* 389–400.

Shallice, T. (1982). Specific impairments of planning. *Philosophical Transactions of the Royal Society, London, B298,* 199–209.

Shiffrin, R. M., & Schneider, W. (1977). Controlled and automatic human information processing: II. Perceptual learning, automatic attending, and a general theory. *Psychological Review, 84,* 127–190.

Sparr, S. A., Jay, M., Drislane, F. W., & Venna, N. (1991). A historic case of visual agnosia revisited after 40 Years. *Brain, 114,* 789–800.

Stiles, W. S. (1978). *Mechanisms of colour vision.* London: Academic Press.

Stoerig, P., & Cowey, A. (1991). Increment-threshold spectral sensitivity in blindsight. Evidence for colour opponency. *Brain, 114,* 1487–1512.

Tallis, R. (1991). A critique of neuromythology. In R. Tallis & H. Robinson (Eds.), *The pursuit of mind* (pp. 86–109). Manchester: Carcanet.

Thomson, D. M. (1986). Face recognition: More than a feeling of familiarity? In H. D. Ellis, M. A. Jeeves, F. Newcombe & A. Young (Eds.), *Aspects of face processing* (pp. 118–122). Dordrecht: Martinus Nijhoff.

Torjussen, T. (1976). Residual function in cortically blind hemifields. *Scandinavian Journal of Psychology, 17,* 320–322.

Torjussen, T. (1978). Visual processing in cortically blind hemifields. *Neuropsychologia, 16,* 15–21.

Tranel, D., & Damasio, A. R. (1985). Knowledge without awareness: An autonomic index of facial recognition by prosopagnosics. *Science, 228,* 1453–1454.

Tranel, D., & Damasio, A. R. (1988). Non-conscious face recognition in patients with face agnosia. *Behavioural Brain Research, 30,* 235–249.

Von Monakow, C. (1885). Experimentelle und pathologisch-anatomische Untersuchungen über die Beziehungen der sogenannten Sehsphäre zu den infracorticalen Opticuscentren und zum N. opticus. *Archiv für Psychiatrie und Nervenkrankheiten, 16,* 151–199.

Warrington, E. K., & Weiskrantz, L. (1968). New method of testing long-term retention with special reference to amnesic patients. *Nature, 217,* 972–974.

Warrington, E. K., & Weiskrantz, L. (1970). Amnesia: Consolidation or retrieval? *Nature, 228,* 628–630.

Weiskrantz, L. (1980). Varieties of residual experience. *Quarterly Journal of Experimental Psychology, 32,* 365–386.

Weiskrantz, L. (1986). *Blindsight: A case study and implications.* Oxford: Oxford University Press.

Weiskrantz, L. (1987). Residual vision in a scotoma: A follow-up study of "form" discrimination. *Brain, 110,* 77–92.

Weiskrantz, L. (1990). The Ferrier lecture, 1989. Outlooks for blindsight: Explicit methodologies for implicit processes. *Proceedings of the Royal Society, London, B239,* 247–278.

Weiskrantz, L., Warrington, E. K., Sanders, M. D., & Marshall, J. (1974). Visual capacity in the hemianopic field following a restricted occipital ablation. *Brain, 97,* 709–728.

Wright, S., Young, A. W., & Hellawell, D. J. (1993). Sequential Cotard and Capgras delusions. *British Journal of Clinical Psychology, 32,* 345–349.

Young, A. W. (1993). Recognising friends and acquaintances. In G. M. Davies & R. Logie (Eds.), *Memory in everyday life.* (pp. 325–350). Amsterdam: North Holland.

Young, A. W., & Bruce, V. (1991). Perceptual categories and the computation of "grandmother". *European Journal of Cognitive Psychology, 3,* 5–49.

Young, A. W., & de Haan, E. H. F. (1988). Boundaries of covert recognition in prosopagnosia. *Cognitive Neuropsychology, 5,* 317–336.

Young, A. W., & de Haan, E. H. F. (1990). Impairments of visual awareness. *Mind & Language, 5,* 29–48.

Young, A. W., & de Haan, E. H. F. (1992). Face recognition and awareness after brain injury. In A. D. Milner & M. D. Rugg (Eds.), *The neuropsychology of consciousness* (pp. 69–90). London: Academic Press.

Young, A. W., de Haan, E. H. F., & Newcombe, F. (1990). Unawareness of impaired face recognition. *Brain and Cognition, 14,* 1–18.

Young, A. W., de Haan, E. H. F., Newcombe, F., & Hay, D. C. (1990). Facial neglect. *Neuropsychologia, 28,* 391–415.

Young, A. W., & Ellis, H. D. (1989). Semantic processing. In A. W. Young & H. D. Ellis (Eds.), *Handbook of research on face processing* (pp. 235–262). Amsterdam: North Holland.

Young, A. W., Hay, D. C., & Ellis, A. W. (1985). The faces that launched a thousand slips: Everyday difficulties and errors in recognizing people. *British Journal of Psychology, 76,* 495–523.

Young, A. W., Hellawell, D., & de Haan, E. H. F. (1988). Cross-domain semantic priming in normal subjects and a prosopagnosic patient. *Quarterly Journal of Experimental Psychology, 40A,* 561–580.

Young, A. W., Hellawell, D. J., & Welch, J. (1992). Neglect and visual recognition. *Brain, 115,* 51–71.

Young, A. W., Newcombe, F., & Ellis, A. W. (1991). Different impairments contribute to neglect dyslexia. *Cognitive Neuropsychology, 8,* 177–191.

Young, A. W., Reid, I., Wright, S., & Hellawell, D. J. (1993). Face-processing impairments and the Capgras delusion. *British Journal of Psychiatry, 162,* 695–698.

Young, A. W., Robertson, I. H., Hellawell, D. J., de Pauw, K. W., & Pentland, B. (1992). Cotard delusion after brain injury. *Psychological Medicine, 22,* 799–804.

Zihl, J. (1981). Recovery of visual functions in patients with cerebral blindness: Effect of specific practice with saccadic localization. *Experimental Brain Research, 44,* 159–169.

Zihl, J., Tretter, F., & Singer, W. (1980). Phasic electrodermal responses after visual stimulation in the cortically blind hemifield. *Behavioural Brain Research, 1,* 197–203.

Zihl, J., & von Cramon, D. (1979). Restitution of visual function in patients with cerebral blindness. *Journal of Neurology, Neurosurgery, and Psychiatry, 42,* 312–322.

IV THE FUTURE OF CONSCIOUSNESS

Antti Revonsuo
University of Turku, Finland

WILL WE GET AN EXPLANATION FOR CONSCIOUSNESS?

It would be most fascinating to catch a glimpse of the future course of science and philosophy. Will we make progress in solving the problem of consciousness, or will the pendulum swing back and forth, as it has been doing for several centuries, consciousness popping up every now and then as a central problem, attacked and found insoluble, then again forgotten or ignored until discovered anew? Philosophers of science usually track scientific theories and discoveries throughout history in order to reveal the hidden patterns of the march of science. But in the case of consciousness, we unfortunately do not possess the wisdom of hindsight. Instead, we are forced to fabricate the philosophy of future science.

Philosophers of science are concerned with such questions as what the aims of science are, whether or not science is progressive, and if it is, how it is possible to measure such progress. Some of them (e.g., Laudan, 1977) propose that the goal of science is the resolution or clarification of problems. Theories ought to provide clarifying answers to baffling problems, showing that what is the case is somehow intelligible and predictable. In Laudan's terms, "Anything about the natural world which strikes us as odd, or otherwise in need of explanation, constitutes an empirical problem" (p. 15). Now there are few things that strike us as less odd than that "technicolour phenomenology arises from soggy gray matter" (McGinn, 1989, p. 349); that the crawling caterpillars and stagnant chrysalises of the brain are magically transformed into the brilliant flying butterflies and moths of conscious experience.

The time is ripe to proceed from mere bafflement towards attempted solutions, that is, theories and models of consciousness. However, as is typical of unsolved empirical problems, it is utterly obscure *whose* problem, precisely, the problem of consciousness in fact is. Which branch of science actually ought to take charge of the mission? From which domain of research should we expect empirical results and relevant theoretical insights? Presently, the division of labor is in a seriously confused state when it comes to consciousness. Prima facie, consciousness seems to belong to the domain of psychology. But psychology was in the grip of behaviorism and psychiatry, accordingly, in that of psychoanalytic theory. Even now that behaviorism is dead and psychoanalysis gradually declining to a historical curiosity, it is hard to come by any serious psychological discussion concerning consciousness. Instead, we have a wealth of other credos by means of which attempts are made to conquer the mountain: Theories and models of consciousness stretch from biological, neurobiological, and neuropsychological all the way to cognitive, computational, and even "neurophilosophical." Furthermore, some think that nothing less than a revision at the level of quantum mechanics is called for and that quantum theory can leap over physics, chemistry, and biology, directly reaching the realm of consciousness. As if this were not enough, some claim that all conceivable theories of consciousness are, in principle, doomed from the start, because they commit false existential presuppositions. That is, all theories attempting to explain consciousness in some terms or other necessarily go astray because, in fact, consciousness does not even exist, and a theory of consciousness consequently is futile.

It is presently very unclear within which domain of scientific inquiry the problem of consciousness should fall, and whether or not it should fall within any domain at all. Perhaps most of the theories claiming their share of consciousness exemplify some of the branches of cognitive science. But if we hope for a resolution from that path, we must face another restraint. From the point of view of the philosophy of science, it is not at all obvious which kind of a science cognitive science strives to be. If it is taken to be just a label referring to practical collaboration between psychology, artificial intelligence, philosophy, linguistics, and brain research, we do not get very far, because then there is no separate theoretical domain that could be named "cognitive" and to which we should deliver both the credit and the blame for explanatory successes and failures. There would still be no one really in charge. If, on the other hand, we accept the strict notion of cognitive science, entailing that there is a unique domain of study for it under which conscious phenomena fall, then we ought to be able strictly to explicate the basic categories of this science and their relationships to other levels of natural phenomena. One would certainly like to know what those basic categories might be—the manipulation of symbolic representations, operations on those representations, computations? Pylyshyn (1989), for example, defended the view that cognition "is literally a species of computing, carried out in a particular type of biological mechanism" (p. 52). Churchland and Sejnowski (1992) believed that "brains are computational in nature" (p. ix) and that "nervous systems are themselves naturally evolved computers" (p. 7).

If the concepts of "computation," "representation," and "biological computer" are indeed the defining notions of cognitive inquiry, then it seems that cognitive science suffers from an exceptionally weak theoretical basis. Which kinds of entities are computations, representations, and computers? Are they natural phenomena? Do they form "natural kinds"? How are they empirically discovered and distinguished from other sorts of natural phenomena? Now, it seems questionable to assume that those terms name any natural processes at all. For example, computers are those physical systems that we decide to *treat as* computers: Sticks and stones can be used as sundials to "compute" time; pieces of plastic put together as a slide rule may be considered to be a look-up table computer. Churchland and Sejnowski (1992, p. 65) recognized this interest-relative dimension: "We count something as a computer because, and only when, its inputs and outputs can usefully and systematically be interpreted as representing the ordered pairs of some function that interests us."

Can the basic notions of cognitive science thus be shown to be dependent on human interests and on the world already constructed by the human mind? If so, then those concepts cannot possibly be an adequate basis for a natural science whose purpose it is to explain the existence of the very same human consciousness already presupposed by its basic concepts. If a preexisting mind is needed to decide which items in the natural world count as computations and computers, then cognitive science explanations of consciousness amount to circularity. Worse still, if computers and computations are not natural kinds, they are not adequate categories of analysis for any natural science approach to consciousness. Natural sciences are involved with discovering and explaining natural phenomena, that is, phenomena that exist independently of human interests and viewpoints (Searle, 1992). Consciousness certainly seems to be a natural enough phenomenon, but the basic notions of cognitive science do not seem to equal, much less explain, consciousness.

In sum, from the viewpoint of the philosophy of science, explaining consciousness is far from a straightforward matter. Although Sperry

(1987) proposed that the replacement of behaviorism by cognitivism was no less than "the consciousness revolution" in psychological science, it now seems that major reformations might still be called for before consciousness is adequately tamed. A not so far-fetched premonition insinuates that consciousness might prove to be a fatal test case for the adequacy of cognitivism. Perhaps the problem of consciousness forces out some fundamental conceptual inconsistencies in the basics of cognitive science, which went unnoticed so far. After all, the fact of consciousness will not go away. To date, many a theoretical Goliath has fallen in the face of that fact.

In chapter 9, Bernard J. Baars and James Newman focus on the borderline between the psychological and the neurobiological, seeking for a unification of theories at these two levels. Among the various natural sciences, there are hierarchical interconnections, and it is generally considered rational e.g., for the biologist to utilize chemical concepts when referring to organic microstructures. A theory of consciousness that makes absolutely no mention of the brain should then raise our doubts. However, cognitive science, at least in its classical form, is exactly like that: The level of neurobiological inquiry is merely compatible with but not positively relevant to theorizing at the cognitive level—one level of exploration entails nothing at all about the other. Now Baars' global workspace theory of conscious experience (see Baars, chapter 7, this volume) clearly has a positively relevant relationship to inquiries at the level of brain neurophysiology. Brain function seems to be, in many cases, organized in a way we might expect on the basis of the global workspace theory. Much work needs to be done with the details, but Baars and Newman certainly have made a promising start. After all, it would have been quite within the bounds of possibility for them to discover that neurophysiology is merely compatible with the global workspace theory, or that neurophysiology entails the implausibility of Baars' theory, or even that the two levels of description are inconsistent with each other—that neurophysiology entails at least a partial negation of the global workspace theory. Things did not turn out that way, so perhaps we are entitled to suspect that there might be a grain of truth in this approach.

The concept of consciousness clearly is a part of the folk-psychological framework, and philosophers of science have recently had violent debates about the fate of this everyman's science in the jaws of scientific progress. Everybody uses folk psychology to understand the behavior of fellow humans and even of themselves. If a person walks into the kitchen but, after arriving there, is utterly confused about the reason for coming, he probably starts to think that he must have *wanted* something from there and that he *believed* that it can be found in the kitchen. Beliefs, desires, and the rest of our common sense psychological vocabulary are the cornerstones of folk psychology. Some philosophers think that the whole explanatory framework should and will go down the drain: There are in reality no such things to which folk-psychological concepts refer. The implication of this view is that also the concept of consciousness ought to be tossed onto the rubbish heap. But, as we noted previously, the fact of consciousness does not disappear even if talk about it is abandoned: We must have learned at least that lesson concerning the blind spots of behaviorism. In chapter 10, Raimo Tuomela presents a philosopher's viewpoint of folk psychology. He argues that although folk-psychological concepts must undoubtedly be revised, it does not imply a wholesale ontological elimination of consciousness and other folk concepts.

In chapter 11, Antti Revonsuo launches the search for a unified science of consciousness. It seems that the multiplicity of various theories, models, and approaches is not only an innocent enrichment for the field but also a potentially damaging or even lethal condition that might inadvertently lead to persistent bedlam in consciousness research, gradually disgracing the scientific approach to consciousness altogether. "Let all the flowers bloom" is a lofty principle, but unfortunately the weeds may shadow and eventually smother the flowers if one has no idea which plants are the ones possibly bearing fruit. Revonsuo proposes that only a consensus concerning the fundamental character of consciousness can guide future empirical research: the nature of consciousness should be made as clear for cognitive science and psychology as the nature of life has been made for biology. Laudan (1977) declared that the first and essential acid test for any theory is whether or not it provides satisfactory solutions to important and interesting problems. In appraising the merits of scientific theories, it is more important to ask whether they constitute adequate answers to significant questions than it is to ask whether they are justifiable within the framework of contemporary epistemology. Revonsuo evaluates current theories and models of consciousness by scrutinizing how they account for two important questions any theory of consciousness ought to answer clearly and unambiguously. These questions he calls "the ontological problem" or "What, basically, are conscious phenomena?" and "the binding problem" or "What mechanism makes it possible for us to experience the world and objects in it as united and coherent, although the brain mechanisms in charge of those experiences are widely scattered throughout the brain?" Are there any theories that give plausible answers to both of these questions? If there are several, do they represent compatible or rival approaches? Do some theories ignore these questions altogether or provide preposterous answers that cannot be incorporated into a larger scientific world view? Such an inquisition might show the way to the dawning of a future science of consciousness.

REFERENCES

Churchland, P. S., & Sejnowski, T. J. (1992). *The computational brain*. Cambridge, MA: MIT Press.

Laudan, L. (1977). *Progress and its problems*. London: Routledge & Kegan Paul.

McGinn, C. (1989). Can we solve the mind–body problem? *Mind, 98,* 349–366.

Pylyshyn, Z. W. (1989). Computing in cognitive science. In M. I. Posner (Ed.), *Foundations of cognitive science* (pp. 49–92). Cambribge, MA: MIT Press.

Searle, J. R. (1992). *The rediscovery of the mind*. Cambridge, MA: MIT Press.

Sperry, R. W. (1987). Structure and significance of the consciousness revolution. *The Journal of Mind and Behavior, 8,* 37–66.

9 A Neurobiological Interpretation of Global Workspace Theory

Bernard J. Baars
The Wright Institute, Berkeley

James Newman
Denver, Colorado

The global workspace (GW) theory described in chapter 7 (this volume) has a natural neurophysiological interpretation. We can perform a neural contrastive analysis of conscious versus unconscious events to pinpoint brain structures and processes that are clearly related to consciousness as such. The most obvious examples are well-known structures whose lesioning abolishes conscious wakefulness: the brainstem reticular formation, the reticular nucleus of the thalamus, and the nonspecific spray of neurons that emerges from the intralaminar nucleus of the thalamus to "activate" the cortex (creating the distinctive electrophysiological signature of waking consciousness). Further, studies of blindsight from damage to the primary visual projection area (V1) make a persuasive case that the contents of a conscious visual experience depend on an intact area V1. Similar results have been reported for the somatosensory primary projection area and the auditory primary projection area. This system of neural structures can indeed be interpreted as a high-level global workspace. In addition to the lesioning evidence, they appear to fit two general properties of global workspace systems, namely competition for input to the GW and widespread dissemination of its output (Baars, 1993; Newman & Baars, 1994). Recent PET scan studies of novel versus automatic perceptual-motor skills show very widespread high glucose utilization when the task is novel and therefore more conscious followed by a dramatic drop in metabolic activity as it becomes automatic, precisely as would be expected from GW theory.

Two major points made in chapter 7 (this volume) are basic to this chapter:

1. We can specify a reliable set of constraints that an adequate theory of consciousness must explain simply by finding contrastive pairs of similar mental events, one of which is conscious and the other not. For example, our memory of arriving here today is currently represented in memory and *un*conscious; we can retrieve parts of it at will, making them conscious, and the original experience was conscious as well. Both the conscious and unconscious cases involve input representation of the same event, differing only in respect to our consciousness of the representation. An adequate theory of consciousness must account for this contrast as well as dozens of others (viz., Baars, 1983, 1988).
2. The set of contrastive pairs describing the typical capabilities of conscious and unconscious processes suggest a particular system-architecture for the nervous system. This architecture, called a "global workspace in a distributed system of specialized processors," has been explored by cognitive scientists for almost 20 years without, however, a systematic attempt to apply it to the issue of consciousness. Such an effort is very fruitful; indeed, numerous contrastive pairs of facts about consciousness can be accommodated in an expanded GW theory. This chapter explores how GW theory points to a set of neural structures that seem to underlie at least perceptual experience.

The neurobiological literature on consciousness and attention has been building for a number of years, really since the 1940s, when the reticular formation of the brainstem that "wakes up" a sleeping or comatose cortex, was discovered by Moruzzi and Magoun (1949). A great deal of important research was done in the 1950s and 1960s on the reticular activating system that can be taken to include not only the brainstem reticular formation (RF) but also the reticular nucleus of the thalamus (NR) and the spray of axons emerging from the intralaminar thalamus to spread to the entire neocortex. One can set up a contrastive analysis table for these major brain-structures involved in consciousness, because with them people can maintain waking consciousness, and without them people and animals go into coma. These contrasts concern the state of waking consciousness rather than some particular contents of consciousness, contrary to the contrastive tables presented previously (Baars, chapter 7, this volume). (Neural content contrasts are discussed later.)

Neuroanatomical and physiological criticism of this work, and especially of its association with consciousness, emerged early. Critics tended to deny that the reticular formation and other structures had anything to do with consciousness—a position that seems untenable given the simple fact that lesions to these structures abolish the waking conscious state. With greater justification, they pointed out that axons emerging from thalamus to cortex were not as "diffuse" as they were first believed to be. The trend in neuroanatomy and physiology was toward smaller and more specific studies of nuclei, small networks, and single neurons. The reticular-thalamic complex was more complicated at this finer level than early investigators realized. It is now known to have many different nuclei; the critics argued that it could therefore not be the "generalized arousal" system that many had suggested.

Such criticisms succeeded in tempering and partly discrediting the first wave of research on the reticular-thalamic complex in the 1950s and 1960s. By 1980 however, in a book called *The Reticular Formation Revisited* (Hobson & Brazier, 1980), a number of neuroscientists were prepared to reconsider the role of the reticular-thalamic system in conscious wakefulness and selective attention. This renewed interest was further supported by an improved understanding of the reticular nucleus of the thalamus (Scheibel, 1980; Skinner & Yingling, 1977) and by a growing understanding of the entire cortex as a massively parallel, distributed collection of specialized processors, most of which are unconscious at any given moment.

One use of theory is to help organize and simplify an otherwise overwhelmingly complex set of research findings. Starting several years ago, we began to try to organize the current evidence regarding the reticular-thalamic activating system in terms of global workspace theory. GW theory had not been developed to deal with neurophysiology, and we had some doubts about the whole enterprise. For example, the current theory does not have a natural way to deal with one of the great anatomical and functional features of the brain, the left-right division of cortex, subcortex, and major sensory and motor pathways.

However, there were some striking coincidences that made a GW approach intriguing. For example, there is very strong neurobiological evidence for parallel distributed unconscious processors, in cortex, for example. As pointed out previously, numerous anatomical structures in the brain serve very specific functions (e.g., Geschwind, 1979; Luria, 1980; and many others). Most parts of the nervous system seem to operate at the same time as other components and to a degree independently from the others. Under these circumstances it is natural to view the brain as a parallel distributed system, and by now many interpreters of brain function have done so. Recently, progress has been made in understanding the workings of parallel distributed systems, and this knowledge has applied to a number of topics (e.g., Rumelhart et al., 1986). A number of similar developments have had the effect of making parallel distributed views of brain functioning the new common wisdom in cognitive neuroscience.

Without a doubt the most extensively studied distributed system in the brain is the cerebral cortex. Mountcastle (1978) viewed the cortex as a collection of specialized distributed processors. He also described patterns of connectivity between these "unit modules" that are not what one might naively expect. Although the gross anatomy of the cortex suggests a "feltwork," or continuous sheet, of convoluted gray matter covering the entire surface of the cerebrum, at the microscopic level this "sheet" turns out to be a great mosaic consisting of millions of functionally discrete columns of cells.

What constitutes a specialized processor? Suppose we speak of a "language processor"—do we need a separate functional processor for syntax, for the lexicon, or even for some particular part of the lexicon? Do the functions of specific areas in the cortex correspond to these traditional subdivisions of language processing? To answer these questions satisfactorily requires looking beyond the specialized functions of any particular module or processor to the way in which columns, modules and processors are organized recursively. That is, a processor may consist of some combination of smaller subprocessors and, in turn, may itself be part of a structured coalition of other processors (much like our earlier analogy of groups of experts). Many of these combinations may appear to be structurally hierarchical, as in the case of perceptual input systems, going from more peripheral to more central, increasingly integrative units. The advantage of viewing hierarchies as recursively organized coalitions is that such coalitions are decomposable into different levels, units, and integrative mechanisms. Thus, there is no "fundamental unit," but rather many units at different levels that may be decomposed and reorganized in different ways to perform different tasks.

An example may help to illustrate this point. When we leap on a bicycle, we would like to access numerous systems—visual, balance, motor control—in a coordinated coalition that behaves for the time being as a single functional processor. But when the left pedal malfunctions, we want to decompose this high-level processor so that we can get our left foot to correct for this unexpected occurrence. Further, when we get off the bicycle, we want to be able to use the same visual, vestibular, and motor systems in a different way to walk or run. The whole coalition of processors must now recognize itself, even though the components of the system remain essentially the same.

The recursive modular organization of specialized processors does not make things easier from a theoretical point of view, but it is consistent with what is known from the careful analysis of cases of localized damage to the neocortex, as in the various subtypes of aphasia. With a lesion to one area of the cortex, large components of speech can drop out, as when a patient loses nearly all capacity for verbal expression (oral and written) in Broca's aphasia, with relative sparing of receptive language functions. In other cases, localized damage results in only very specialized functions being impaired. For example, in alexia without agraphia the lesion results in a selective incapacity to read with nearly normal understanding and expression of oral speech. Even writing and spelling are spared, although the patient may not be able to read what he has written. Works on clinical neuropsychology provide many more illustrations of the localization of specialized cognitive functions in the human cortex (e.g., Geschwind, 1979).

Altogether, the evidence is strong that the cortex and other brain structures can be treated as large collections of specialized systems that can operate in parallel. We now turn to the claim that other parts of the brain can be viewed as a global workspace, or its functional equivalent. What might be the neural substrate of such a global workspace? Table 9.1 shows a set of contrasts between similar conscious and unconscious structures and processes.

Consider the massive behavioral and neural difference between someone who is in deep sleep or coma compared to one who is awake and alert. No alien space visitor could fail to observe that vertebrates, including humans, cease any purposeful movement for a third of the earthly day and that human beings can report no memory of inner or outer events that occur in deep sleep. This obvious behavioral cycle corresponds exactly to massive electrophysiological changes throughout the brain, as the fast, irregular, and low-amplitude activity characteristic of wakefulness is replaced by the slow, regular, high-amplitude waves of deep sleep. It also happens to correspond exactly to reports of conscious experience.

TABLE 9.1.
Neurobiological Contrasts

Conscious	*Unconscious*
State of Consciousness	**State of Unconsciousness**
Intact RF, NR, and nonspecific thalamocortical fibers	Lesions in any of these structures
Contents of Consciousness	**Unconscious "contents"**
Perceptual cortical activity	Preperceptual cortical activity (Libet, 1981)
Visual conscious experience with intact area V1	Blindsight due to damaged area V1: some representation without reported conscious experience
Non-habituated ERPs throughout brain	ERPs limited to sensory input system (Thatcher & John, 1977)
Widespread high metabolic activity with novel computer game (Haier et al., 1992)	Lower metabolic activity in the same subject's brain after practice
Attended channel activity	Non-attended channel activity

The neuroanatomical structures necessary for the state of consciousness are well known: They include the brainstem reticular formation, the outer shell of the thalamus (its reticular nucleus), and the fountain of nonspecific neurons that projects widely from the thalamus to all parts of the cortex (Fig. 9.1). Absent any of these structures, humans and animals lapse into coma. In contrast, any part of the great cerebral cortex and most subcortical structures can be lost without impairing conscious wakefulness. Thus, the state of consciousness has well-established behavioral, neurophysiological, and anatomical correlates.

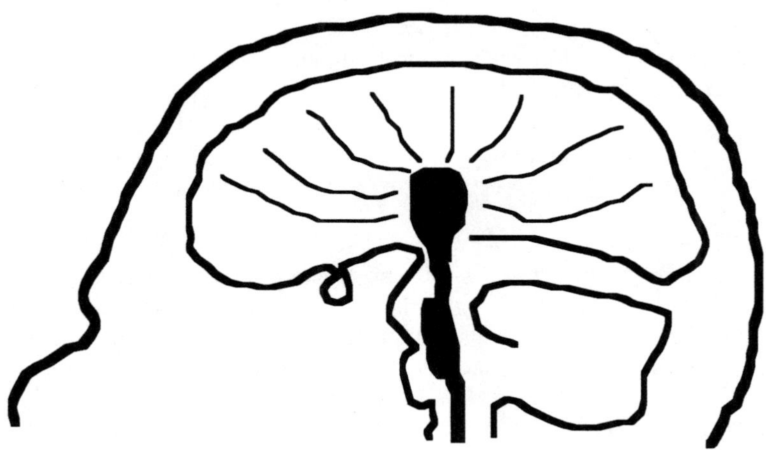

FIG. 9.1. The extended reticular-thalamic system: A neural global workspace? Brain structures most closely associated with conscious experience include the reticular formation of the brain stem and midbrain, the outer shell of the thalamus, and the set of neurons projecting upward diffusely from the thalamus to the cerebral cortex. Together these structures can be labeled the Extended Reticular-Thalamic Activating System (ERTAS) because stimulation of a number of neurons in this system causes cortical activation. The ERTAS as a whole can be interpreted as a functional global workspace or perhaps as an increasingly global set of workspaces. ERTAS has many properties reminiscent of global workspaces including connections in both input and output with all sensory and motor systems, and with almost all areas of the brain; known competition between different inputs and the possibility of "global broadcasting" of information through the diffuse thalamic projection system. It is possible that other brain structures also serve the function of global broadcasting including the corpus callosum connecting the two hemispheres and the tertiary corticocortical fibers that connect distant parts of the cortex. Turn Fig. 9.1 upside down, and it looks like a GW diagram.

An adequate theory of consciousness must clearly deal with the contents of consciousness as well, because when awake, we can experience an indefinite number of distinct conscious contents including percepts and images in all sensory modalities, memories, fantasies about possible futures, and the like. In addition, we can gain conscious access to a vast range of abstract concepts, intentions, expectations, and actions. As discussed in the chapter introducing global workspace theory (Baars, chapter 7, this volume), there is much evidence to show that stimulus representations can be conscious or unconscious. Libet (1981) showed that cortical activity in the somatosensory area is present a few tens of milliseconds after stimulation; but the cortical activity may not become conscious until as long as 500 milliseconds of sensory stimulation. Cortical event-related potentials seem to indicate that the nervous system is not passively representing the unconscious somatosensory stimulus; major cognitive events go on between 40 and 400 milliseconds post stimulus.

TWO MAJOR ASPECTS OF GLOBAL WORKSPACES: INPUT COMPETITION AND OUTPUT DISTRIBUTION

Specialized input processors compete for access to a global workspace. Once a processor gains access, it can disseminate its message broadly throughout the society of specialists. We know of course that different potential conscious contents, such as two streams of speech, compete for access to consciousness. Once such access is gained, humans can do an extraordinary range of things with the input, possibly because of global broadcasting of conscious contents. Thus, input competition and output dissemination characterize a broadcasting facility, much like a television station, newspaper, or scientific podium. Do the anatomical structures associated with the state of consciousness show any such functional pattern? Somewhat to our surprise, they do indeed.

Evidence for Access Competition

Two of the neural structures discussed previously that are associated with conscious wakefulness—in the sense that their absence results in coma—show competition between different input modalities. They are the brainstem reticular formation (RF) (Hobson & Brazier, 1980; Magoun, 1962) and the nucleus reticularis (NR) of the thalamus (Scheibel, 1980). The RF is believed to modulate the activity of many higher-level neural structures, notably the neocortex, whereas the NR is believed to control thalamic "gatelets" that can facilitate or inhibit relay synapses for sensory tracts on their way to the cortex. The NR specifically shows competition between inputs that correspond to the same location in body-centered space.

There is a problem. As candidates for a possible GW, the problem with both of these systems is that their output bandwidth does not seem to be large enough to carry the information for a conscious visual scene, for example. Further, phenomena like blindsight suggest that primary cortical projection areas are needed for conscious perceptual experiences.

In sum, there is considerable evidence for competition between potentially conscious perceptual contents for access to neural structures such as the RF, NR, and primary sensory projection areas, which are strongly associated with consciousness on independent grounds.

Global Workspace Output Is Also Characterized by Widespread Dissemination of Conscious Information

Given competition for input to the neural correlate of a global workspace, what about the output function? Some well-established phenomena suggest widespread, global distribution of conscious perceptual information in the nervous system.

1. General measures of cerebral activity show that even very specialized tasks, when they are novel and presumably require more conscious involvement, show widespread activity in the cortex.

If cerebral activity during specialized tasks consisted solely in the activation of a few specialized processors, we might expect cortical activation patterns to form distinctive "patch works" with significant areas of the cortex being essentially quiescent. This is seldom the case, however. Studies of hemispheric specialization employing both electrophysiological (EEG) and metabolic (e.g., cerebral blood flow and PET scans) measures of cortical activity consistently show widespread activation of large parts of *both* hemispheres during the performance of specialized tasks, such as speaking or silent reading. Although levels of activation in areas known to be specialized for particular cognitive functions are generally greater, differences in activity levels between homologous areas in the two hemispheres seldom exceed a few percentage points (Donchin, Kutas, & McCarthy, 1977; Springer & Deutsch, 1981). Lassen, Ingvar, and Skinhoj (1978) reported, "a general increase in blood flow [which] seems to be distinctly related to the subject's effort in performing the task, because it is absent when simple tasks are performed but is evident when the subject shows signs of struggling with a difficult problem" (p. 70). Clearly mentally effortful tasks require a more deliberate, conscious, and nonautomatic attempt by the subject to perform the task. True automaticity tends to be acheived only in tasks that are entirely predictable, so that they are experienced as effortless (Shiffrin & Schneider, 1977). Haier and his colleagues also showed that a beginner playing a computer game (TETRIS) shows very widespread metabolic activity as measured by PET scans, whereas the same individual shows lower and more localized activity after practice and thus less conscious involvement (Haier et al., 1992). Although there are competing explanations for each of these phenomena, together they make a substantial case for widespread dissemination of brain activity related to consciousness of the activity.

2. The orienting response, closely associated with conscious surprise as novelty, is known to involve every major division of the nervous system.

Another commonplace experience, whose neural basis has been extensively studied, is our tendency to be drawn to surprising or novel events in our environment. The obvious purpose of this orienting response (OR) is to focus consciousness on novel or potentially dangerous stimuli. The OR is probably the most widespread reflexive response of the nervous system. It desynchronizes EEG alpha (resting) activity throughout the cortex, dilates or contracts blood vessels all over the head and body, changes the conductivity of the skin, causes orienting of the external sensory organs to the source of novel stimulation, triggers changes in autonomic functions such as heart rate and peristalsis, rapid pupillary dilation and so on. All of these responses need not be produced by a globally broadcast message, but the fact that they are so extraordinarily widespread, both anatomically and functionally, suggests that something of that sort may be occurring.

3. Apparently any central nervous system (CNS) event can come under voluntary control, at least temporarily, using conscious biofeedback techniques.

It is not generally emphasized enough that biofeedback training always involves *conscious* feedback information. To gain control over alpha waves in the EEG, we pay attention to a tone or light corresponding to the increased alpha activity; to gain control over muscular activity, we play back a conscious click for each peak of electrical activity in the relevant muscle, and so on. In terms of global workspace theory, to establish biofeedback control over previously uncontrolled activity in the body, we must first "broadcast" the feedback in some way. Presumably, the neuronal systems controlling the alpha waves and other biofeedback responses behave as independent by distributed processors able to receive the global message and act upon the coincidence of the global information with its own activity.

Under such conditions, people can gain at least temporary control over an extremely wide range of physiological activities with surprising speed. Biofeedback control of single neurons has been established in animals in the hippocampus, thalamus, hypothalamus, ventral reticular formation, and preoptic nuclei (Olds & Hirano, 1969). Larger populations of neurons can also be controlled, including alpha waves in the EEG, activity in the sensory-motor cortex, evoked potentials, and activity in the lateral geniculate nucleus (Chase, 1974). In the muscle system, single motor units can come under conscious control with half an hour or so of training, and with further biofeedback training subjects can learn to play drumrolls on single spinal motor units (Basmajian, 1979). (Spinal motor units consist of only two neurons.)

As Buchwald (1974) wrote, the ability to train control of neuronal structures "is so ubiquitous a phenomenon that there seems to be no form of CNS activity (single unit, evoked potential, or EEG) or part of the brain that is immune to it" (p. 15).

This is what we would expect if conscious feedback were made available throughout the brain, and local distributed processors "decided" whether to respond to it. We may draw an analogy between biofeedback training and trying to locate a child lost in a very large city. At first, it makes sense to search for the child around home or school, in a local and systematic fashion. But if the child cannot be found this way, it makes sense to broadcast a message to all the inhabitants of the city to which only those citizens who found it relevant would respond. The message is global, but only those with pertinent knowledge will respond to it.

In sum, it is difficult to imagine an account of the power and generality of biofeedback training without something like global broadcasting in the nervous system.

4. Event-related potential (ERP) studies show that prior to habituation, conscious perceptual input is widely distributed in the brain.

There is some direct neurophysiological evidence for global broadcasting associated with consciousness. E. R. John and his colleagues published a series of experiments using event-related potentials (ERPs) to trace the neural activity evoked by repeated visual or auditory trains of stimulation, such as a series of bright flashes or loud clicks (John, 1976; Thatcher & John, 1977). Typically an awake, unanesthetized cat, with a number of electrodes implanted throughout its brain, is presented with a series of light flashes. Electrical activity is monitored via the implanted electrodes and averaged in a way that is time locked to the stimuli, so that random activity can be averaged out. In this way, remarkably simple and clean event-related traces emerge from the seeming randomness and complexity of ordinary EEG activity.

The main finding of interest here is that ERPs evoked by the visual flashes are initially recorded at all the electrodes—far beyond the specialized visual pathways. At this point, we can assume that the cat is having an orienting response to the novel stimuli and is thus, to the best of our knowledge, consciously attending to the light flashes (Sokolov, 1963). But as the same stimulus is repeated, habituation of orienting takes place. ERPs in response to the repeated stimuli never disappear completely but become more and more localized—until finally they are limited only to the classical visual pathways.

The results of these experiments are strikingly in accord with our expectations. According to our model, prior to habituation, the stimulus information is conscious and globally distributed. But after habituation or orienting, it ceases to be conscious and becomes confined to only those areas of the brain that are specialized for visual functions. Only these specialized input processors are now involved in analyzing the stimulus.

How global is the brain's global workspace? On the basis of the evidence cited previously, we answer, "Very global indeed." This is suggested especially by the literature on biofeedback and the orienting response as well as the neurophysiological evidence concerning the global distribution of event-related potentials prior to habituation. This is a strong claim to make, but one that seems consistent with current knowledge of the anatomy and physiology of the reticular-thalamic activating system. This system, detailed below, is widely considered to be necessary for the *state* of conscious wakefulness, itself a necessary condition for conscious perceptual experience.

The evidence can therefore be summarized as follows:

1. RF and NR activity seem to be a necessary but not sufficient for conscious experience;
2. Stimulus representation in primary sensory projection areas also seems to be necessary but not sufficient for conscious perceptual experience (Weiskrantz, 1986).

The simplest hypothesis is that *both* components are necessary and sufficient to support conscious perceptual experience. If the RF and NR can select between different inputs, so that the selected input is fed to the primary projection area for vision (in our example), there is plenty of bandwidth for broadcasting information from V1 to the rest of the brain.

The primary visual projection area in the cortex, area V1, is not generally thought to contain a very rich representation of the visual world. However, V1 receives numerous projections from "higher" visual projection areas, of which there may be as many as 25. Therefore it may be that the dynamic content of V1 is not limited to the rather primitive features of line orientation that are known to be represented there, but may include much more sophisticated "downward" influences representing complex objects, faces, color, movement, scene recognition, and the like. Alternatively, another one of the early projection areas, singly or in combination, may provide such a rich source of combined multileveled visual representation that can then be disseminated widely to the remainder of the cortex.

However, if the primary visual projection are in the cortex (V1) is to be the visual global workspace, broadcasting is not likely to occur instantaneously. Rather, there may be a feedback loop between the primary thalamic visual relay nucleus and area V1. It is noteworthy in this respect that most of the projection fibers between thalamus and V1 are running "the wrong way," going downward from cortex to thalamus, perfectly situated to support a two-way feedback loop.

Once there is a stable feedback loop, dissemination of the contents of V1 may occur through very rich connections between area V1 and all other areas of the cortex. In this scheme the nonspecific projections from the reticular thalamus may serve to start the looping of visual information, and to suppress competing auditory and somatosensory loops via the gating system of the reticular nucleus. Voluntary ability to look at a visual scene, rather than listen to speech or feel bodily pain may occur through frontal cortex control of the reticular gating system (see Newman & Baars, 1994, for one detailed proposal).

Thus, for consciousness of a visual object, like a coffee cup, one plausible account is that a thalamocortical feedback loop, along the lines of Edelman's "re-entrant signaling loop" (Edelman, 1989) may be necessary in order to establish a specific cortical focus corresponding to the stimulus (Kinsbourne, 1988). Once the loop is established, information from primary visual cortex can be broadcast via massive connections from striate cortex to subcortical mechanisms via the thalamus and to other cortical areas via wide-band connections such as the arcuate fasciculi, tertiary cortical tracts, commissural tracts to the contralateral hemisphere and spread of activity through the "feltwork" of Layer 1 of the cortex (Newman & Baars, 1994).

Notice that other cortical visual areas can feed back to area V1 via retrograde flow of information. Thus, visual cortical information would not be limited to the V1 columns for line orientation but would be able to incorporate higher-level analyses of the conscious visual scene including higher-level object and event representations. Thus, V1 would act as a GW for dissemination of global information elsewhere in the brain and also as a topographical "binding" surface in which both higher and lower levels of visual analysis could be integrated (see Crick & Koch, 1990).

How does the dissemination, or global broadcasting, occur? If GW is presumed to be area V1, the primary projection area, there is plenty of bandwidth for broadcasting conscious contents from corticocortical and corticofugal tracts. That is:

1. There are two-way thalamocortical projections between the specific thalamic nuclei of the sensory pathways, gated by NR;
2. There are massive corticocortical fiber bundles, including the arcuate fasciculi, tertiary corticocortical connections to frontal and temporal lobes from striate cortex, and connections to the contralateral hemisphere;
3. Newman & Baars (1994) suggest, along the lines of von der Malsburg and Singer (1988), that the horizontal feltwork of Layer 1 of the cortex allows for lateral spreading of information from a cortical focus in the primary sensory projection areas to other parts of the cortex.

For example, if the primary visual projection area V1 acts as an input processor, it may gain downward-flowing access to the diffuse thalamic projection system and recruit other parts of cortex via nuclei of the thalamus including nucleus reticularis, intrathalamic nuclei with their diffuse cortical connections, and the specific sensory and other relay nuclei. This information may then disseminate from V1 via the great swaths of neurons flowing to all parts of cortex and subcortex. Such a system could be guided by means of the "gatelets" of the nucleus reticularis discussed by Scheibel (1980) that in turn may be controlled by the attentional system in parietal and frontal cortex, as suggested by Posner and others (e.g., Posner, 1992). Thus, access to consciousness may be controlled by attentional systems that may be voluntary, as in frontal control, or involuntary, in more posterior areas.

Rather than having a one-shot broadcast of conscious percept, we may expect a rapid cyclical recruitment of the most active and significant current input by a cycle of activity starting from the visual primary cortical projection area, looping downward into the thalamus, then upward again, using both indirect and direct pathways to other parts of cortex, pulling in the massive corticocortical and corticofugal pathways described previously (see Damasio, 1989; Edelman, 1989).

ACKNOWLEDGMENTS

The first author is grateful to Professor David Galin (UCSF) for several years of enlightening conversations on human brain functions and to Dr. Katharine McGovern, whose comments were again a great help in presenting and writing this paper. Comments on related work by F. H. C. Crick, Charles Yingling, and Ted Melnechuk were most valuable.

REFERENCES

Baars, B. J. (1983). Conscious contents provide the nervous system with coherent, global information. In R. Davidson, G. Schwartz, & D. Shapiro (Eds.), *Consciousness and self-regulation* (pp. 45–76). New York: Plenum.

Baars, B. J. (1988). *A cognitive theory of consciousness.* New York: Cambridge University Press.

Baars, B. J. (1993). How does a serial, integrated and very limited stream of consciousness emerge from a nervous system that is mostly unconscious, distributed, and of enormous capacity? In G. R. Brock & J. Marsh (Eds.), *CIBA symposium on experimental and theoretical studies of consciousness* (pp. 282–290). London: Wiley & Sons.

Basmajian, J. V. (1979). *Biofeedback: Principles and practice for the clinician.* Baltimore, MD: Williams & Williams.

Buchwald, J. S. (1974). Operant conditioning of brain activity. An overview. In M. H. Chase (Ed.), *Operant conditioning of brain activity.* Los Angeles: University of California Press.

Chase, M. H. (Ed.) (1974). *Operant conditioning of brain activity.* Los Angeles: University of California Press.

Crick, F. H. C., & Koch, C. (1990). Towards a neurobiological theory of consciousness. *Seminars in Neurosciences, 2,* 263–275.

Damasio, A. R. (1989). Time-locked multiregional retroactivation: A systems-level proposal for the neural substrates of recall and recognition. *Cognition, 33,* 25–62.

Donchin, E., Kutas, M., & McCarthy, G. (1977). Electrocortical indices of cortical specialization. In S. Harnad (Ed.), *Laterlization in the nervous system* (pp. 57–93). New York: Academic Press.

Edelman, G. (1989). *The remembered present: A biological theory of consciousness.* New York: Basic Books.

Geschwind, N. (1979). Specializations of the human brain. *Scientific American, 241(3),* 180–201.

Haier, R. J., Sieger, J. R., MacLachlan, A., Soderling, E., Lottenberg, S., & Buchsbaum, M. (1992). Regional glucose metabolic changes after learning a complex visuospatial/motor task: A positron emission tomographic study. *Brain Research, 570,* 134–143.

Hobson, J. A., & Brazier, M. A. B. (Eds.). (1980). *The reticular formation revisited: Specifying function for a non-specific system.* New York: Raven Press.

John, E. R. (1976). A model of consciousness. In G. Schwartz & D. Shapiro (Eds.), *Consciousness and self-regulation* (pp. 6–50). New York: Plenum Press.

Kinsbourne, M. (1988). Integrated field theory of consciousness. In A. J. Marcel & E. Bisiach (Eds.), *Consciousness in contemporary science.* London: Oxford University Press.

Lassen, N. A., Ingvar, D. H., & Skinhoj, E. (1978). Brain function and blood flow. *Scientific American, 239(4),* 62–71.

Libet, B. (1981). Timing of cerebral processes relative to concomitant conscious experiences in man. In G. Adam, I. Meszaros, & E. I. Banyai (Eds.), *Advances in physiological science* (pp. 235–242). Oxford: Pergamon Press.

Luria, A. R. (1980). *Higher cortical functions in man* (2nd edition). New York: Basic Books.

Magoun, H. W. (1962). *The waking brain* (2nd ed.). Springfield, IL: Thomas.

Moruzzi, G., & Magoun, H. W. (1949). Brain stem reticular formation and activation of the EEG. *EEG and Clinical Neurophysiology, 1,* 455–473.

Mountcastle, V. B. (1978). An organizing principle for cerebral function: The unit module and the distributed system. In G. M. Edelman & V. B. Mountcastle (Eds.), *The mindful brain* (pp. 75–143). Cambridge, MA: MIT Press.

Newman, J., & Baars, B. J. (1994). A neural attentional model for access to consciousness: A global workspace perspective. *Concepts in Neuroscience, 2(3).*

Olds, J., & Hirano, T. (1969). Conditioned responses of hippocampal and other neurons. *EEG and Clinical Neurophysiology, 26,* 159–166.

Posner, M. I. (1992). Attention as a Cognitive and Neural System. *Current Directions in Psychological Science, 11,* 11–14.

Rumelhart, D. E., McClelland, J. L., & the PDP research group (Eds.). (1986). *Parallel distributed processing: Explorations in the microstructure of cognition.* Cambridge, MA: MIT Press.

Scheibel, A. B. (1980). Anatomical and physiological substrates of arousal. In J. A. Hobson & M. A. Brazier (Eds.), *The reticular formation revisited: Specifying function for a non-specific system* (pp. 42–59). New York: Raven Press.

Shiffrin, R. M., & Schneider, W. (1977). Controlled and automatic of human information processing II: Perceptual learning, automatic attending, and a general theory. *Psychological Review, 84,* 127–190.

Skinner, J. E., & Yingling, C. D. (1977). Reconsideration of the cerebral mechanisms underlying selective attention and slow potential shifts: Central gating mechanisms that regulate event-related potentials and behavior. In J. E. Desmedt (Ed.), *Attention, voluntary contraction and event-related potentials: Progress in clinical neurophysiology, Vol. 1* (pp. 30–69).

Sokolov, E. N. (1963). *Perception and the orienting reflex.* New York: MacMillan.

Springer, S. & Deutsch, G. (1981). *Right brain, left brain.* San Francisco: Freeman.

Thatcher, R. W., & John, E. R. (1977). *Foundations of cognitive processes.* Hillsdale, NJ: Lawrence Erlbaum Associates.

von der Malsburg, C., & Singer, W. (1988). Principles of cortical network organization. In P. Rakic & W. Singer (Eds.), *Neurobiology of the neocortex* (pp. 69–95). Berlin: Wiley.

Weiskrantz, L. (1986). *Blindsight: A case study and implications.* Oxford: Clarendon Press.

10 The Fate of Folk Psychology

Raimo Tuomela
University of Helsinki, Finland

FOLK PSYCHOLOGY

A couple of decades ago much was being said about the reduction of psychology and the other social sciences to neurophysiology. Today this reduction, at least in the received sense of deductive theory-reduction by means of type-type connections, has come to seem a very unlikely possibility. Instead, one now speaks of the correction and, possibly, elimination of psychology, or one opposes such views. In this chapter, the issues in this debate are surveyed and commented on mainly from the point of view of scientific realism. To the extent I am developing and defending a position on the relevant ontological and epistemological problems, it is, vaguely speaking, that folk psychology needs correction without a wholesale elimination of its basic categories. The new ideas and arguments to be found in this chapter involve mainly a clarification of the nature of folk psychology and an argument for correctivism.

In recent philosophical literature, it has become common to refer to doctrines or belief systems that people share in the form of a common folklore or common cultural heritage. Such doctrines, to the extent they are not refined by science, may be called folk doctrines. Here one may also speak of the manifest image (as contrasted with the scientific image) of the world and mean by it the kind of views that members of the Western culture and Western social collectives are commonly taught by their parents and other comparable educators. Accordingly, there is folk psychology (FP), folk medicine, folk meteorology, and so on. Folk psychology can be construed as a framework (with both conceptual and factual elements) in terms of which people think of themselves and others. Folk psychology in this sense—which has been called the "framework of agency"—goes back as far as Aristotle. However, I would like to mention the important modern contributions by the philosopher Sellars (1968), the social psychologist Heider (1958), and the social psychological attribution theory based primarily on Heider's views. Especially the ideas of Sellars have affected the thinking of current advocates of eliminativism (see Churchland, 1979, 1989). It should be said, however, that the framework of agency really is a philosopher's construction rather than an ordinary layperson's view of people's mental lives. A more detailed discussion is warranted.

There are those who emphasize the theory-like character of folk psychology and those who emphasize its other functions (e.g., its normative character and its use in role taking and simulation). A typical example of the former, the "theory-theory" view, is seen in the following quote from Churchland (1981): "Each of us understands others, as well as we do, because we share a tacit command of an integrated body of lore concerning the lawlike relations holding among external circumstances, internal states, and overt behavior. Given its nature and functions, this body of lore may quite aptly be called folk psychology" (p. 69). I shall here characterize folk psychology in a sense that covers not only this theory-theory view but that also is able to accommodate the other views mentioned. In my view, folk psychology consists of a mutually accepted or believed set of representations that the collective in question happens to have at some moment whether those representations are true or not. This set of representations or sentences (in a natural language or possibly in some kind of language of thought) will be the "body of lore" in question.

Speaking about a social collective, folk psychology can be viewed as follows: It is assumed that there is in this collective a mutual belief (or a mutually shared presupposition) to the effect that these representations are true or acceptable. Some of these representations are rule-of-thumb generalizations or maxims about propositional attitudes and actions (e.g., about how actions come about because of strong desires in suitable doxastic conditions). That there indeed exist such generalizations that are mutually accepted by Western societies is clearly the case: Consider what people say about each other's mental states, what fictional literature says, what social psychologists say (e.g., think of attribution theory), and what many philosophers say about the nature of our common sense psychology. To be sure, some of the representations are about normative matters—what one should do or think in various circumstances; some others relate to speech acts such as warnings or commands. Furthermore, some of the representations can at least tentatively be assumed to be about nonpropositional psychological matters (e.g., direct perception and many mental skills such as empathizing). Thus, it can be a mutual belief in this collective that its full-fledged members can take the roles of or simulate others in various situations (e.g., of a person in distress or one who wants to have a certain object) without this role taking having to be based on any generalizations, at least explicitly and knowingly.

It is assumed that some of the representations in the content of folk psychology are psychological generalizations and that folk psychology therefore has theory-like aspects. This is, of course, an empirical assumption: From a conceptual point of view, things might have turned out differently. Nevertheless, given the truth of this assumption, one can still—at least for the time being—ontologically be either a realist or an instrumentalist about what the representations are about; and the same goes for the mutual belief itself. I concentrate on the realist construal of the propositional attitudes and the mutual belief, because the current dispute in the literature about the fate of folk psychology commonly makes this assumption. Often such realism relies on the view that true or best-explaining theories determine what there is in the world and that folk theory is taken by its proponents to be such a theory. (For instrumentalists the ontological question does not arise; they have only to deal with the relevant conceptual and epistemic problems.)

Given the the previous information, one can, at least tentatively proceed with the assumptions that the body of lore to be called folk psychology includes (a) a set of common sense platitudes and (b) a thesis about the existence of propositional attitudes, actions, and the like and that there is a mutual belief and acceptance of (a) and (b) in the social collective in question.[1] Construing folk psychology in this way, one can meaningfully ask whether the mutual belief in question is true, viz., whether human beings are persons in the sense of their satisfying—or even approximately satisfying—(a) and (b).

As seen, one can in principle be either a realist or an instrumentalist —or hold some kind of intermediate position—concerning the basic categories of folk psychology. Under both interpretations things may go well, in the sense that the best-explaining neuroscientific theory can yield corrected type-type correlations between mental types and neurophysiological types, or they may go badly if one gets no such nice connections. In the latter case, if things turn out very badly, one can arrive at a situation in which strictly speaking there are no beliefs, no intentions, no actions, no norms, no institutions, and so on, although there still might be some kind belief-like states. In the following discussion such replacing states (or counterpart states) are called *Ersatz*-states. Presently of course, it is not known what will be the case, but one may at least hope for and even reasonably expect the former situation—a "Panglossian reconstruction." It is a kind of reconciliation of the falseness of FP with the noneliminability (even in principle) of the basic conceptual categories of psychology. Thus, under this Panglossian construal, there would, for instance, still be beliefs, although many statements now in FP taken to be true of them would turn out to be false. The position that is defended later, although being naturalistic, it does not attempt to reduce intentional notions to nonintentional ones and is compatible with the acceptance of a suitably corrected FP.

FOLK PSYCHOLOGY AND THE TRADITIONAL MIND–BODY PROBLEM

When discussing the nature and fate of folk psychology, one cannot avoid taking up recalcitrant problems related to the mind–body problem as it has been conceived traditionally. Following a brief introductory survey of some of the central problems (based on Tuomela, 1985), I consider the recent discussion related to the elimination and correction of folk psychology, and—hardly surprisingly—many old friends, to be surveyed in this section, are then encountered.

In medias res, thinkers from Plato and Aristotle on have assigned to human minds various properties that can be regarded or are regarded as nonmaterial. For example, Rorty (1979) demonstrated that the mind–body problem has several times during the course of two millennia adopted a new guise; the criteria of matter have varied accordingly. To get a picture of the complex mind–body problem, I present a representative list of the central features that have been used in descriptions of the mental aspects of people (cf. Feigl, 1967, p. 29; Rorty, 1979, p. 35):

1. incorrigible self-awareness
2. ability to sense qualities
3. ability to exist separately from body
4. nonspatiality
5. ability to be aware of universals
6. intentionality
7. ability to use language
8. ability to act freely
9. ability to be "one of us" and to act morally

Feature 1 (self-awareness) has often been regarded as incorrigible (cf. "Even if my perception of that table may be nonveridical, yet surely it is an indubitable fact that it seems to me that there is a table over there"). Feature 2 relates to sensory *qualia* like the smell of a rose or the bright redness of a new car. Also qualities that are not at least primarily sensory might be considered here, for example, my feeling of satisfaction about having solved a difficult problem. Features 1 and 2 can be regarded as Cartesian mental features and so can obviously Features 3 and 4. Understanding 1–4 from the Cartesian perspective, one can say that they have to do with mental substance and its properties. For Descartes that means thoughts, experiences, feelings, and so on that occur in the stream of consciousness. These occurrent episodes have the property that they would exist even if one thought that everything else (such as matter) were annihilated or otherwise reduced away. Features 3 (ability to exist separately from body) and 4 (nonspatiality) are of course connected with the idea of such mental substance, whether it consists of changing processes ("the ghost in the machine", as for Descartes) or of some immutable substance (cf. the *noûs* of the Greek and the soul of Christianity). But also Feature 1, incorrigible awareness of these experiential states and processes, is central. In its essentials it involves the epistemic version of the "Myth of the Given," the idea that "the categorial structure of the world—if it has a categorial structure—imposes itself on the mind as a seal imposes an image on melted wax" (Sellars, 1981, p. 12). In other words, if an item (object, state, etc.) either in the mind (e.g., a headache) or in external reality (e.g., a stone) has a certain feature (e.g., the mental item is a painful headache or the physical item is a brown stone), then immediate awareness of this item gives incorrigible knowledge of it. Sellars directed effective criticisms against this myth (also cf. Tuomela, 1985).[2]

Many discussants (e.g., Jackson, 1986; Kripke, 1972; Nagel, 1974; Searle, 1992) have taken the sensing of qualia, Feature 2 (e.g., the smell of a rose or the quality of a headache) to be a (or the) constitutive feature of the mental and the central problem for materialism. There have also been forceful replies to this challenge by proponents of functionalist mind–body theories (cf. Lycan, 1987) and others (cf. Churchland, 1989). However, partly on common sense grounds and partly because of the very idea of a feeling and sensing person that one is supposed to deal with here, I regard this aspect of the mental as important. I take for granted the platitude that sentences purportedly ascribing mental states and events can be applied and often do apply (truthfully or at least in an instrumentally useful way) to human beings. This fact by itself should and need not be taken to commit the ascriber to incommunicable subjective mental entities and states: the use of sentences ascribing mental states and the truth-makers or satisfiers of those sentences should and can be kept analytically separate. Following the arguments by Wittgenstein, Sellars, and many others, I assume that the primary conceptual and epistemic basis (but without strict logical connections) for such mental ascriptions is public social practices. In particular, I would like to emphasize overt "Rylean" notions such as overt actions and (actional as well as nonactional) thinkings-out-loud.[3] Mental states and processes are kinds of theoretical entities construed on this basis. It can be said that human beings can and do have intentional mental states as well as mental states with special phenomenal characters. The question of the precise nature of the satisfiers is not fixed a priori (this is part of what "theoreticity" here involves). Therefore, the existence of mental states thus construed is not an obstacle to the scientific view of man but instead something that simply must be considered when giving a scientific account of human beings as feeling and sensing persons. Even if sensory and other mental qualities (or, more precisely, putative ascriptions of such features) ultimately could not be accounted for in a strictly materialistic way—understanding matter as current neural science understands it—materialism in a weaker sense is nevertheless defensible.[4] This weaker kind of materialism—or naturalism or monism, if you prefer—still excludes the possibility of the existence of phenomenal objects (as opposed to states with phenomenal features). It possibly satisfies even the debated *scientia mensura* thesis, according to which the method of science is decisive.

It seems that the notions of a purely mental object and substance do not have too many advocates in current philosophy. In any case, no one has been able to give a clear and coherent characterization—free of category mistakes—of the notion of purely mental substance or object (cf. Rorty, 1979; Ryle, 1948; Wittgenstein, 1953). Although this problem need not be discussed here, let me say generally that mental activities can be understood without postulating ontically significant separate mental entities such as headaches and after-images—entities conceptually formed out of mental activities. (For recent arguments to this effect, see Lycan, 1987.)

Features 5–7 (ability to be aware of universals, intentionality, and ability to use language) have traditionally been associated with questions of human intelligence. Thus, the problem of the ability to have direct awareness of universals and the ability to grasp universals was pivotal for both Plato and Aristotle (cf. again the Myth of the Given). Note specifically that Aristotle's *noûs*, intelligence or reason, was in a sense a faculty separate from body (cf. Feature 3) but nevertheless in many respects totally non-Cartesian (see Rorty, 1979). But if universals can be analyzed to be basically linguistic entities, the ability to grasp universals would hardly seem to be a mysterious metaphysical problem. I operate in this paper on the assumption that this (Sellarsian) view is basically correct. (Cf. Seibt, 1990, for a recent discussion and defense.)

It is of some interest here to note that Greek philosophy—Aristotle's own philosophy in particular—did not recognize the mind–body problem in the Cartesian form, that is, in the form of purely mental entities that have special phenomenal qualities and that are in some sense self-verifying. For Aristotle, the counterparts of these Cartesian entities (e.g., headaches) belonged squarely to the body and were therefore material (see Matson, 1966); this I find a plausible view to hold. Perhaps the most central features of the mind–body problem, that is, Features 1–4 (incorrigible self-awareness, ability to sense qualities, ability to exist separately from body, and nonspatiality) are historically rather recent and derive, for the most part, from Descartes.

As to intentionality, it can basically be regarded as representation and aboutness (or semantic directedness), presupposing that there are "representers" and "representeds." Full-blown intentionality is something that human languages are capable of and that purposive and norm-based human use of language involves. The intentionality of mental states is to be understood in close analogy with the intentionality of such "languagings," actional and nonactional. We explicate the intentionality of languagings in terms of the notion of a linguistic rule in a Sellarsian fashion, linguistic rules including not only intralinguistic ones but also world-language and language-world rules. Thus it can be said that intentionality relates to metalinguistic categories (such as "means," "denotes," "is about"), and they are to be elucidated in terms of norms (or rules) of language and in the case of Sellars, ultimately intelligent, patterned languagings.[5] When intentionality is characterized along these lines, it is not ontically problematic; it does not entail dualism, and it does not ontically add to the notion of intelligent languaging and the notion of rule-obeying, for instance. That linguistic rules or norms can be taken to be involved entails that intentionality is also a social category, norms being social. By analogy, propositional or sentential mental states (e.g., wanting and believing) can also be thought to be intentional. Furthermore, one can also speak of lesser forms of intentionality (e.g., in the case of animals high up on the evolutionary scale) based on poorer representational means than human beings have.

However, there are other views about intentionality, and some of them regard intentional mental states as the locus of intentionality and explicate the intentionality of languagings on that basis. It is not possible here to consider arguments *pro et con*. I simply assume that Wittgenstein and Sellars were on the right track: The intentionality of linguistic items is the central notion on which the intentionality of mental items relies. Which beings are intentional depends on the contingent issue of capability of languaging or approximation to languaging.

Features 8 (ability to act freely) and 9 (ability to be "one of us" and to act morally) refer to the category of personhood. It can be argued by following Sellarsian lines of thought that these features should be analyzed by reference to prescriptive discourse. No ontological problems arise here, if one can assume—as I do in the company of many other philosophers—that prescriptions involve no ontological commitments and that thus there are no intrinsically normative states of affairs (cf. Tuomela, 1985).

As these admittedly scant remarks indicate, a materialist who accepts the Kantian (or Kantian-like) distinction *quaestio facti – quaestio juris* (or order of being – conceptual order) taking only the former dimension to be ontically significant, and who, even in a more important sense, analyzes universals in linguistic terms, seems to steer around the troubles involved in the mind–body problem. This Sellarsian (or, rather, almost-Sellarsian) approach to the mind–body problem results in a kind of emergent materialist position of mind as "highly organized matter" that if not correct, is at least a promising candidate for further exploration. Its general leading idea is that man is a material system that interacts in various ways with its environment and that is emergent relative to neurophysiological microentities and conglomerations of them as presently conceived. However, this emergence cannot be strong emergence but may be something like supervenience (relative to the brain states and possibly some naturalistic environmental states). As a technical term of philosophy, supervenience involves the idea that brain states determine mental processes. Thus, it can be thought that the emergence is not causal emergence, viz., whatever causal powers mental states and events have accrue to them from their nonmental (material) features. Given that the identity of entities must ultimately be explicable by means of their causal powers, there is not a strong kind ontic emergence here. (But supervenience does not reduce mental states to brain states in terms of necessary and sufficient conditions thus preserving what could be called the mental level.) Although there is not ontic emergence, at least not in a strong sense, one nevertheless can speak, in the crucial cases, of a kind of conceptual emergence and of the pragmatic ability at the moment to operate with mental concepts rather than material ones in the crucial cases.

The emergent features to be accounted for include such subjective sensory features as smells and colors as well as such social phenomenal features as the feeling of shame or that of liking somebody. The scientific account of these features must not essentially change them but must keep to the topic. This matter-energy system can be described by material (or more generally, nonpsychological) as well as by psychological predicates (and in some sense needs both kinds of predicates for its satisfactory description). Future neuropsychology might well merge with psychology in such a manner that it nevertheless contains predicates counterpart to today's psychological predicates, at least to most of them.

Trying to be more precise, one wants to apply ascriptions to human beings by using intentional and intensional mental words, and the primary conceptual and epistemic basis for mental ascriptions are public Rylean notions such as overt actions and thinkings-out-loud. A central question now is which kinds of objects, states, and events must exist for that to be possible. My tentative answer has been that only material or naturalistic objects, states, and processes need to be supposed to make true mental sentences (ascriptions). But I have not claimed that the mental words or their counterpart words in some future theory might not be needed. If mental predicates or their counterparts indeed will be needed, my position will be dualistic with respect to descriptions—to the extent such a descriptive dichotomy of predicates can ultimately be made sense of. The present program need not involve anything like a reduction of mentalistic sentences or theories. (Nominalistically speaking, only naturalistic objects, perhaps characterized in terms that are not materialistic in a strict sense, are needed.) Intentionality is not reducible or need not be regarded as reducible in this program, but it is important to see that it does not commit one to more than languaging (the obeying of linguistic rules in a less than a fully actional sense; recall note 5).

As long as one rejects the linguistic Myth of the Given, saying that there are no a priori privileged conceptual systems, the application of psychological predicates does not make one's ontology dualistic by creating an irreducible category of mental substance or something of the kind. In other words, man, as a matter-energy system, is in part describable if one uses present or future psychological predicates and, without change of topic, is not adequately describable as a person without them or their counterparts within the scientific image. It is important to see that because one does not directly connect language with ontology, one has a view that does not, and need not, reduce away mentalistic (intentional and other) descriptions and that yet is naturalistic and compatible with ontological materialism (or monism, if you prefer).

ARGUMENTS FOR AND AGAINST ELIMINATION OR CORRECTION

Given the previous introduction, I next consider the recent philosophical discussion on the problem of the status of FP. This discussion takes up many of the mentioned aspects of the traditional mind–body problem; it would be surprising if it did not. The following list states some central arguments that have been presented for and against the replacement (or at least correction) of the basic categories of FP:

A. Arguments for the Elimination (or at Least Correction) of FP:

1. When considered as a theory or a theory-like entity, FP seems clearly false. It is—or so the argument takes it to be—shallow and narrow, something to be compared with rules of thumb. Arguments to this effect have been presented by Nisbett and Ross (1980), Kahneman, Slovic, & Tversky (1982), Churchland (1979, 1989), Stich (1983), and many others.
2. Ascriptions of propositional attributes (e.g., belief) are essentially vague and observer-relative (see Needham, 1972; and especially, Stich, 1983).
3. The intentional vocabulary and ontology of FP resist incorporation into the growing synthesis formed by the other sciences including such neighboring sciences as biology and neurophysiology (Churchland, 1979).
4. FP fails to account for the mental processes of children, the mentally ill, and of primitive and exotic peoples (Churchland, 1979; Stich, 1983).
5. If connectionism is right, then propositional attitudes are eliminable (Ramsey, Stich, & Garon, 1990). There is some reason to think that the antecedent is or can be made true.

B. Arguments Against the Elimination of FP:

1. Folk theory works reliably in many (indeed most) cases. For instance, people tend to fulfill their promises and usually obey social norms, and predictable behavior ensues. People can generally rely on such regularities when planning action and acting. So one can say that FP is reliable for action,

interaction, and the maintenance and renewal of social institutions.

2. The elimination of the basic categories of FP, such as belief and action, is not possible because the very doctrine or thesis of elimination cannot be coherently and truly stated and believed. For instance, the eliminativist should be taken to believe that there are no beliefs. But if there are no beliefs, the eliminativist cannot have the mentioned belief. So the assumption that the eliminativist has that belief leads to a conclusion contradicting that assumption, from which the impossibility of eliminativism follows.
3. Persons are necessarily intentional beings, and intentionality is an uneliminable category, characterizable only in terms of FP.
4. Which beliefs one ascribes to someone depends noncausally on social context. Accordingly, Burge (1986) has argued that a person's physical state may remain constant while the social environment changes, and in such a situation, the person's mental states, that is, beliefs, may change and vary as a function of changes in the social environment. Psychological states thus are uneliminably anchored in the social environment.
5. There are subjective phenomenal features, "feels" or qualia that cannot be accounted for without folk psychology. For instance, the smell of coffee or—to take a nonsensory example—the feeling of shame, according to the present argument, cannot be explicated or explained without the conceptual resources of folk psychology and—as is sometimes added—without ontic dualism.
6. FP cannot be eliminated because it is needed for conceptualizing the forms of life that make agents capable of moral and esthetic judgment, make possible moral and legal responsibility, communication, self-expression, and so on.
7. There is the pragmatic difficulty involved in learning a new conceptual framework for our ordinary affairs as well as for conducting scientific research. For instance, the categories of observation (perceiving, noticing, etc.) would change, and this would pose a pragmatic problem for testing our theories—not to speak of more mundane activities requiring the conceptualization of sensory inputs and outputs.

It seems that all of the previous arguments can be found in the literature on the topic, although perhaps not quite as I have formulated them (in addition to the previous references, see Garfield, 1988). Next I briefly discuss these arguments especially from the point of view of ontology. To begin with, it should be pointed out that eliminativism (cf. Churchland's position) typically presupposes that the world can in principle be described and conceptualized in various ways. This means the denial of the Myth of the Given referred to earlier (see note 2). The rejection of the ontological, linguistic, and epistemic versions of the Myth of the Given serves to make correctivism (and at least to some extent eliminativism) possible in principle. It should be noted, however, that the correction (or elimination) of metalinguistic categories and thus intentionality—as I construe this notion—is not touched. I assume this much without further argument. What is more problematic is what is pragmatically possible, viz., possible for real flesh-and-blood people in actual life.

Argument A.1 is especially concerned with people's mistakes in reasoning and contradictions in their belief systems and other similar faults in the rationality of their mental processes. It is not possible here to go into any detail, except to note that it is somewhat difficult to judge the situation (even when the correctness of the relevant experimental results is assumed). For which reasons is it justified to give up the category of belief as opposed to making corrections in what is assumed of beliefs? Under which conditions is it justified to make the latter kinds of corrections as opposed to saying that the subjects are stupid, misinformed, irrational, or the like? A fuller discussion is not possible here; the reader is referred to the mentioned literature. To conclude, at least tentatively, concerning Arguments A.1 and A.2, at least some corrections in what is said or can be said about people's beliefs may be called for—noting that FP is not unambiguous as to what it presently claims about them. The observer-relativity of belief, convincingly argued for by Stich, seems to pose a clear difficulty for the employment of the common sense notion of belief in the context of scientific psychology.

Arguments A.3 and A.4 give additional support to the claim about the unsuitability of the common sense mentalistic notions for scientific psychology. Although psychology is perhaps going to need belief-like states, desire-like states, and so on, their family resemblance to the notions of FP may not be high (see Stich, 1983). Stich did leave it as an open possibility that although future cognitive science will not invoke the concepts of common sense psychology, it will nonetheless postulate state types, many of whose tokens turn out to be describable as the belief that *p*, the desire that *p*, and so on.[6] This is a Panglossian scenario. According to Stich's other, more pessimistic scenario, there are no beliefs, for the assumptions of similar "functional architecture" and "modularity" may fail.[7] Argument A.5 specifically claims that connectionism can be regarded as a competing cognitive theory or approach (and not only an approach to clarify the neurophysiological underpinnings of cognition). However, presently it is not clear whether or not connectionism is incompatible with FP, the arguments of Ramsey et al. (1990) notwithstanding. Even if it were incompatible, it is not clear whether connectionistic theories are going to supersede psychological theories correcting FP while basically building on the conceptual resources of FP.

Argument B.1 concerns action rather than mental processes. It is a strong point against the elimination of at least the category of action. As to desire–belief, one and the same action can ensue from different desire–belief constellations depending on circumstances (and the converse seems true, too). Furthermore, rational and reasonable action may ensue from a contradictory or irrational belief–desire constellation (see Nisbett & Ross, 1980, for demonstrations). This may seem to indicate that the notions of belief and desire are up for correction and even elimination much sooner than the notion of action. However, the matter primarily concerns poor reasons rather than the correction or elimination of the notions of belief and desire.

Argument B.2 indicates that if folk notions are eliminated from future science, the elimination must be wholesale and encompass the metalanguage one uses to comment upon this future science and its philosophy. It also makes it plausible to think that such future science must bear a suitable family resemblance to FP or else it would not be dealing at all with the same subject matter. There would thus have to be (possibly connectionistically conceived) Ersatz beliefs, desires, and action, and there would accordingly be Ersatz beliefs about Ersatz beliefs, and so on, not necessarily strictly type-type correlated with their predecessors but rather conceived of in a Panglossian way. Thus, it is not contradictory to say that an eliminativist Ersatz-believes that there are no beliefs in the sense of FP. One can also say here that the intentionality of beliefs is preserved "modulo Ersatz."

It seems that philosophers like Kripke and Searle would be willing to argue along the lines of Argument B.3, but I am not prepared here to give chapter and verse as proof. What can be said against Argument B.3 from the scientific realist viewpoint is this: Human beings have no deep, hidden metaphysically (ontologically) important essence that would set them apart from other real beings, such as apes, amoebas, or even stones (cf. Rorty, 1979). To be sure, it is a categorial principle of the framework of agency that persons necessarily are intentional beings.

But nevertheless human beings are not necessarily intentional: It can be regarded only as a conceptually contingent fact that members of the species *Homo sapiens* are intentional in more or less the sense of the framework of agency. In other words, it is a contingent fact, to the extent it is one, that human beings are persons in the sense of FP. Furthermore, if intentionality is characterized in metalinguistic semantic terms it is not intimately tied to the resources of FP. Human beings clearly seem to be representation-making beings, but those representations might not be precisely what FP takes them to be. Thus Argument B.3 seems definitely too strong. (Below a partial "pragmatic" counterpart to Argument B.3 will be accepted.)

In recent literature especially Burge argued for Argument B.4, and I here basically accept his argument, stressing along with Burge (1986) that the point at bottom is not a linguistic one (see, however, the rebuttal by Bach, 1988). Against the argument that beliefs depend on social contexts, one might nevertheless try to claim that, first, it might be possible to distinguish between the truth conditions of belief ascriptions and the psychological content of belief states. If that were possible, eliminativists could say that they will only substitute neurophysiological content for such psychological content, while letting the truth conditions remain unchanged—at least as long as FP as a whole has not been eliminated. In spite of this, I still take Argument B.4 to give much more ammunition to the conservationist than the eliminativist or even the correctivist.

The qualia problem has been much discussed in the literature. Because comments on it were made in the previous section, I only make two additional remarks. First, the qualia problem is often associated with the existence of a kind subjective or perspectival facts. Thus, it is argued that it is not possible for an observer to feel what I feel when I have, say, lower back pain. A materialist can accept that this feeling is essentially subjective in a trivial sense: you cannot have my (token of) lower back pain. (You cannot perform my particular arm-lifting or be subject to my particular stumbling either; and this has nothing to do with the mind–body problem.) I can also have and know what it is to have lower back pain. To fully know what it is to have a lower back pain, I may have to get a back problem, but at least as long as we are participating in the same form of life and in the same language game, I can not only have lower back pain but also meaningfully communicate with you about lower back pain. All this is as such independent of whether people are material beings or made of, for example, ectoplasm (although I believe that the specific nature of sensory feels does *factually* depend on what people happen to be made of). Anyway, in this sense, there are no essentially subjective (viz., incommunicable) facts. The other alternative is that there are essentially private thoughts and feelings and that people thus, so to speak, are in the private-language business. I do not discuss it here (there is already a voluminous post-Wittgensteinian literature on this subject) except for noting that that question also seems to be independent of the mind–body problem, namely, of what do people consist of. I do, however, assume in this paper that something like the Wittgenstein-Sellars view about the existence of a private language is right, taking this to entail that there are no essentially subjective (viz., incommunicable) facts: All facts are communicable and corrigible in principle. Accordingly, one can say that qualitative mental features also occur in a cognitive context and, indeed, can be said to have a communicable cognitive component. Thus, the qualitative feature of the visually sensed greenness of a birch-leaf is an aspect of a sense-impression or perception that might turn out to be illusory or hallucinatory. There is always a corrigible component present even in internal cases like feeling a pain. The concluding thesis of correctivism later concerns in part this fact.[8]

Arguments B.6 and B.7 take up features of FP that arguably are not at least centrally relevant to its truth or falsehood and therefore to its theory-like aspects and functions.[9]

What can one now conclude from the previous discussion? If none of the previous arguments is regarded as fully decisive alone, the problem of arriving at a conclusion concerning whether or not to eliminate FP or to correct it becomes one of weighing the arguments against each other. Taken together and stressing the open a posteriori nature of the situation, the presented arguments for and against the elimination or at least revision of ordinary mental notions do pose a difficulty for the conservative theorist who wants to keep intact the psychological and social psychological notions of FP when investigating the laws and regularities of the social realm. The arguments also make difficult the wholesale elimination of FP. However, the point about a substantial correction of FP by a theory or science retaining the folk-psychological notions of belief and of other mental states does not seem to be affected by the previous arguments. In that case, people would, under normal circumstances, have beliefs and would act and so on, more or less in the sense of FP. As to Argument B.1, reliability may even be improved, as the new replacing theory or science would obviously provide better explanations and would correct the falsehoods of FP.[10]

AN ARGUMENT FOR FULL-BLOWN CORRECTIVISM

Correctivism can be justified also in a more precise way by using the premises of scientific realism. I next consider an argument based on four familiar premises:

1. Assumption underlying the in-principle-possibility of correction within realism: The ontological, linguistic, and epistemic versions of the Myth of the Given are rejectable.
2. Assumption concerning the determination of ontology within scientific realism: Best-explaining (or true) theories specify what there really is in the world.
3. Assumption about the nature of full-blown intentionality: The intentionality (intentional aboutness) of intentional items (linguistic items, representational mental states, and so on) can in its most central sense be understood on the basis of semantic metalinguistic terms that are ontically neutral with respect to the mind–body problem; thus thoughts, contrary to or possibly contrary to sense impressions, are ontically noncommittal. The understanding involved is only analogical in the case of mental states and episodes. Furthermore, the weaker kinds of intentionality exhibited by nonlinguistic organisms are treated on the basis of their similarity to language-users.
4. Assumption about research interests: The research topic of psychology is human beings as thinking, feeling, and acting socialized persons.

These assumptions warrant the following *Thesis of Correction*: As long as human beings are viewed as thinking, feeling, and acting persons, the correction and even elimination of the basic conceptual categories of FP and of the rule-of-thumb generalizations concerning them is possible as far as ontological, conceptual, and epistemic limitations are concerned, although such changes may not be pragmatically possible.

Assumption 1 says that no conceptual system is privileged: thus one can speak of Ersatz beliefs and, in principle, even of neural states XYZ instead of beliefs. (However, assumption 4 does restrict the topic.)

Assumption 2 makes the method of science the best method for gathering knowledge as opposed to mystical or religious methods or the methods the notorious man of the street uses. The notion of a best-explaining theory is an idealization: It is what rational inquirers using the method of science can achieve when free from human limitations (e.g., physical and psychogical limitations concerning perception, memory, and reasoning). The method of science is meant to have the features of objectivity, criticalness, testability, self-correctiveness, autonomy, and progressiveness. Self-correctiveness deserves special emphasis by virtue of its importance among methods of gathering knowledge (see Tuomela, 1985, esp. chs. 9 & 10). To say that something really exists versus that it does not exist is my shorthand for saying that its existence is based on best-explaining science or it is not. Thus, relying on the present knowledge, which surely falls short of being based on best-explaining science, one can conjecture that although there are chairs and trees in the world, they do not really exist: What really exist are the microphysical counterparts of chairs and trees. This remark need not make the reader's firm grip of his chair any less firm: The doctrine in question is not crazy in the sense of denying the existence of common sense objects in a sense making them somehow imaginary. Rather, the view concerns the right way to carve up the world into causally right bits and pieces, explanatory power of theories being a central criterion here.[11]

Assumption 3 serves as a reminder that intentionality is not an ontically significant category but dependent on linguistic practices and norms for them. In this sense assumption 3 can be regarded as a pragmatic assumption, involving a Sellarsian kind of functionalist view of mental states. Assumption 4 centrally restricts the topic and prevents the wholesale elimination of folk psychology. However, it allows for the use of Ersatz beliefs in the place of ordinary beliefs: Elimination is possible as long as the topic is not changed, and this may involve more than the mere tidying up the edges. It may involve a change of the basic conceptual categories as long as the topic (viz., human beings as thinking, feeling and acting persons) is not forsaken. That the topic must, in such a liberal sense, be kept constant is a kind of conceptual necessity that however, need not be taken to imply ontological constraints. The issue about the topic of psychology has deliberately been formulated in a vague fashion so as to minimize the a priori constraints that are placed on the developing science of psychology.[12]

Obviously the Thesis of Correction says that everything belonging to the topic of human beings as persons is correctable: There are no incorrigible a priori truths about human beings as persons although there are conceptual truths about persons!

ACKNOWLEDGMENT

I wish to thank George Berger for comments.

NOTES

[1] Let me here present some examples of generalizations that FP can be taken to contain. In Tuomela (1977) I considered "meaning-specifying" postulates for the common sense notion of wanting and present the following list, ranging from rather clearly analytic to rather clearly synthetic generalizations about this notion supposedly holding for all full-fledged agents and all want contents (also cf. Audi, 1973):

(W1) Agent A wants that p if and only if , given that A has not been expecting p, if A now realized that p is or will be the case, then A would immediately be disposed to feel satisfaction about this.

(W2) A wants that p if and only if, given that A has been expecting p, if A now realized that p would not be the case, then A would immediately be disposed to feel dissatisfaction about this.

(W3) If A does or would find daydreaming about p pleasant, and A believes that there is at least some considerable probability of p's occurring, then, under favorable conditions, A wants that p.

(W4) A wants that p if and only if for every action X which A has the ability and opportunity to perform, if A believes that his performing X would have at least some considerable probability of bringing about p, then A is disposed to do X, given that A has no stronger competing wants.

(W5) If (a) it would be unpleasant to A to entertain the thought that p is not (or probably is not the case), or the thought that p will not be (or probably will not be) the case, and (b) A believes that there is at least some considerable probability of p's occurring, then, under favorable conditions, A wants that p.

(W6) If A wants that p, then (a) A is disposed to think or daydream about p at least occasionally, and (b) in free conversation A tends to talk, at least occasionally, about p and subjects that he believes are connected with p.

(W7) A wants that p if and only if, under favorable conditions, A has a disposition to want-out-loud that p.

(W8) If A wants that p, then, if A is frustrated in his attempt to bring about p, then A is disposed to develop latent aggression.

For a discussion of folk science see Tuomela (1994), especially chapter 9.

[2] Sellars (1963) discussed the Myth of the Given and takes it to result in a number of objectionable views. In Sellars (1981) the epistemic myth is given the following formulation: If someone is immediately aware of an item that in point of fact (i.e., from the standpoint of "best explanation") has categorial status A, then he will be aware of it as having categorial status A.

In Tuomela (1985) and (1988) I argued that there are the following three objectionable versions of the Myth of the Given, the ontological (MG_o), the epistemic (MG_e), and the linguistic (MG_l):

(MG_o) There is an ontologically given, categorially ready-made real world.

(MG_e) Man can be engaged in nonconceptual but yet cognitive epistemic commerce with the world.

(MG_l) There is an irreplaceable, a priori privileged language (or conceptual framework).

[3] In Sellars (1963) a program—in terms of his mythical character Jones—for the formation of psychological concepts on the basis of public Rylean notions is sketched and defended. See also Tuomela (1977) for an attempt to make the Jones myth more precise.

[4] Recall here the *sensa* postulated by Sellars (1971). They are entities in the scientific image postulated in accordance with the *scientia mensura* thesis (see note 11). In his terminology, they are physical$_1$ (part of the causal order, something required for the best explanation of nonliving or living matter) even if not physical$_2$ (necessary and sufficient for the best explanation of nonliving matter).

[5] In Sellars's (1963) theory of language, meanings are nonrelationally taken to be analyzed in terms of "dot-quotes." Thus, roughly, the metalinguistic semantic statement

"The expression 'bachelor' in English means an unmarried man" is rendered as "The expression 'bachelor' in English is a .unmarried man." in which the dot-quoted expression ".unmarried man." applies to any expression in any language with the same semantical role as "unmarried man" has in English. Semantic roles are specified in terms of following or satisfying rules of language: intralinguistic rules, world-language, and language-world rules.

In Sellars's theory the notion of "pattern-governed behavior" is needed as a notion intermediate between full-blown, mentalistic rule-obeying and (more or less) accidental conforming to a rule. The important thing is that pattern-governed behavior need not be mentalistic in the sense of involving an intention to behave in accordance with the rule. For instance, (Sellarsian) perceiving-out-loud, inferring-out-loud, and expressing emotions are behaviors that are even necessarily pattern-governed, namely, they can never be performed intentionally, with the intention to realize their end-result (I cannot properly intend to see that tree over there or intend to infer a new logical result or feel sorrow for something). Full-fledged language-users ought to follow the rules of language (world-language, intralanguage, and language-world rules) of their community or at least they ought to be disposed to follow them. These "oughts" cash out the central content of Sellars's dot-quotes (concerning representational, such as linguistic, items).

[6] Stich (1983, pp. 224 ff.) investigated the belief state types of FP, the generalizations of FP, and the relation between cognitive science and the social sciences. He claimed, and with justification, that generalizations in FP are like rules of thumb in cooking. Such generalizations are vague, and they cannot be improved considerably; and cognitive science is supposed to explain why and in which respects they are partly true and partly false.

[7] Consider here Needham's (1972) study of belief-like mental states in a variety of societies. He concluded that there is nothing that matches up with our own notion of belief: There is no such thing as believing that *p*.

[8] Lycan (1987, p. 148) presents a list of eight types of qualia problems that are to be found in the literature and that he claimed materialism (or at least his "homuncular" materialism) can handle. Lycan summarized his position on the problem of the subjectivity of the mental in the following way that I can agree with: "Human and other subjects can have functionally or computationally different states that nonetheless home on the same objective state of affairs, either external or internal. But there are no intrinsically subjective or perspectival facts that are either the special objects of self-regarding attitudes or facts of 'what it is like'. There are only states of subjects that both function in a particularly intimate way within those subjects and have the subjects themselves and their others states as inevitable referents. And that, I think, is all there is to 'subjectivity'" (Lycan, 1990, p. 126).

In a book that appeared after the present article had been written Searle (1992) regarded the subjectivity of mental states as an ontological fact. For a relevant discussion and rebuttal of Searle's view, see Dennett (1993).

[9] This claim presupposes that normative or prescriptive discourse—which according to the argument is needed to characterize the framework of agency—does have ontic commitments (see the criticism of this claim in Tuomela, 1985).

[10] To comment on corrective explanation, supposeone is dealing with a situation where T_i and T_{i+1} represent, respectively, a psychological predecessor theory and a successor theory provided by cognitive science. The former theory is supposed to be couched in the terminology of FP, whereas the latter uses concepts of cognitive science, for example, connectivist concepts.

The following two technical schemes of corrective explanation (discussed in Tuomela, 1984, 1991) can now be considered:

I T_{i+1} *correctively*$_1$ *explains* T_i if and only if

(a) there exist an auxiliary hypothesis *H* and a theory T_i^* such that

(i) T_{i+1}&H explains T_i^*, and

(ii) T_i is an approximation of T_i^*;

(b) T_{i+1} is a better explanatory theory than T_i.

II T_{i+1} *correctively*$_2$ *explains* T_i if and only if

(a) there is a translation function *tr* from the conceptual scheme (or language) of T_i into the conceptual scheme (or language) of T_{i+1} and an auxiliary hypothesis *H* and a theory T_i^* (both in the language and conceptual scheme of T_{i+1}) such that

(i) T_{i+1} jointly with *H* explains T_i^*, and

(ii) $tr(T_i)$, viz. the translation of T_i into the conceptual scheme (or language) of T_{i+1}, is an approximation of T_i^*;

(b) T_{i+1} is a better explanatory theory than T_i.

Scheme I concerns smooth scientific reduction by means of correspondence rules. However, it seems rather likely that the relation between FP and cognitive science is not of this kind, recall the arguments for correctivism. Scheme II tries to make precise a notion of corrective explanation in which the reducing theory corrects the reduced theory by explaining what was right and what was wrong in the latter. Reduction here does not yet quite entail elimination.

It is not possible here to discuss Scheme II in detail, but I will briefly present some results from an earlier investigation (see Tuomela, 1991). To begin, an analysis of meanings in terms of linguistic roles (cf. Tuomela, 1985, after Sellars) may be more precise than a truth-condition analysis. For instance, two predicates may have the same truth conditions or satisfiers and still have different use-meanings (think of familiar school examples such as "bachelor" versus "unmarried man", or something more complicated). Assume here (probably counterfactually) that such a finer analysis is available at least in principle.

Scheme II speaks of of a translation function *tr*, defined in precise terms in Tuomela (1991). This translation function is a truth-preserving one: It connects every sentence of the predecessor language with a sentence of the successor language in the following sense: Given that a certain general connection between the relevant models (satisfiers) holds, if the predecessor sentence is true, then so is also its translation.

Given all this, there is the following results:

1. Meaning variance (difference in use-meaning or linguistic role) is compatible with translatability.
2. Comparability on a token-token basis (viz., as it were, the same fact viewed in two different ways) is possible on the thin, extensional level of comparison. This is actually an assumption one makes rather than a result. It is plausible, for unless the predecessor scientist and the successor scientist can agree on the fact that they are studying the same phenomena in a rough extensional (viz. thin) sense and thus are using in part the same measurement apparatuses and are able to "point to" the same thinly understood phenomena (in a suitably thin sense), there is no sense in speaking of them as scientists working somehow in the same field (not to speak of labeling them specifically as predecessor and successor theorists). On the thin level of comparing predecessor and successor theories, it is plausible to think that the models of these theorists' languages can be compared on a token-token basis. Let us denote by *R* the relation so pairing

these respective models. (It is perhaps somewhat idealized to require that all the models can so be paired, but this slight idealization has the virtue of removing some of the slack and indeterminacy in the comparison; cf. point 4.) Technically speaking, this then is the result: The model-theoretic relation *R* holding between two singular states of affairs (represented by models) gives truth-preserving translation if *R* is a so-called "projective relation."

3. Meaning variance is compatible with comparability (more specifically, *R*-comparability, whatever the latter exactly amounts to in real scientific practice). This observation has not properly been argued for previously, but it follows from plausible ideas making use-meaning more fine-grained than meaning explicated in terms of truth-conditions.
4. We may still have indeterminacy of translation in two different ways: (a) for every translation function *tr* there can be many ways of translating any given sentence; (b) there may be several translation functions *tr*. The truth of this point relies on the very setup and the assumptions made; see notes 2 and 11. It is not claimed here that this indeterminacy could not be removed, at least in some cases, by making additional assumptions.

One can now say that, as obviously some correction of FP is needed, both schemes I and II seem to give viable explications of the relevant kind of correction and corrective explanation, and they both are compatible with the nonelimination of FP in an ontic sense.

[11] The idea of scientia mensura is basically the idea that the best-explaining theories constructed in accordance with the method of science (which need not be a priori fully specified) are the measure of what there is in the world. There are, however, stronger and weaker explications of this idea. I discuss some central versions in Tuomela (1985). For the present purposes the following weak version seems to suffice:

(*MSR*) All sentient and nonsentient physical objects have at least the constituents and properties that correspond to the scientific terms needed in the theories that best explain the overall behavior of those objects.

The following Sellarsian version is stronger and, on balance, the most plausible:

(*SSR*) All sentient beings (persons included) and all nonsentient physical objects have precisely the constituents and properties that correspond to the theoretical scientific terms needed in the theories that best explain their overall behavior, where the explanations also cover the sensory features of the world and man's reactions to them.

[12] Assumption 4 can be connected in an obvious way to the scheme of corrective explanation and the translation function *tr* spoken about in note 10. Space does not permit a fuller discussion here.

REFERENCES

Audi, R. (1973). The concept of wanting. *Philosophical Studies, 24,* 1–21.

Bach, K. (1988). Burge's new thought experiment: Back to the drawing room. *The Journal of Philosophy, LXXXV,* 88–97.

Burge, T. (1986). Individualism and psychology. *Philosophical Review, 45,* 3–45.

Churchland, P. (1979). *Scientific realism and the plasticity of mind*. Cambridge: Cambridge University Press.

Churchland, P. (1981). Eliminative materialism and propositional attitudes. *Journal of Philosophy, LXXVIII*, 67–90.

Churchland, P. (1989). *A neurocomputational perspective: The nature of mind and the stucture of science*. Cambridge, MA: MIT Press.

Dennett, D. (1993). Review of Searle (1992). *The Journal of Philosophy, XC*, 193–205.

Feigl, H. (1967). *The "mental" and the "physical"*. Minneapolis: University of Minnesota Press.

Garfield, J. (1988). *Belief in psychology*. Cambridge, MA: MIT Press.

Heider, F. (1958). *The psychology of interpersonal relations*. New York: Wiley.

Jackson, F. (1986). What Mary didn't know. *The Journal of Philosophy, 83*, 291–295.

Kahneman, D., Slovic, P., & Tversky, A. (Eds.). (1982). *Judgment under uncertainty: Heuristics and biases*. Cambridge: Cambridge University Press.

Kripke, S. (1972). Naming and necessity. In D. Davidson & G. Harman (Eds.), *Semantics of natural language* (pp. 253–355). Dordrecht: Reidel.

Lycan, W. (1987). *Consciousness*. Cambridge, MA: MIT Press.

Lycan, W. (1990). What is the "subjectivity" of the mental? *Philosophical Perspectives, 4*, 109–130.

Matson, I. (1966). Why isn't the mind–body problem ancient? In P. Feyerabend & G. Maxwell (Eds.), *Mind, matter, and method* (pp. 92–102). Minneapolis: Minnesota University Press.

Nagel, T. (1974). What is it like to be a bat? *The Philosophical Review, 83*, 435–450.

Needham, R. (1972). *Belief, language, and experience*. Oxford: Blackwell.

Nisbett, R., & Ross, L. (1980). *Human inference: Strategies and shortcomings of social judgment*. Englewood Cliffs, NJ: Prentice-Hall.

Ramsey, W., Stich, S., & Garon, J. (1990). Connectionism, eliminativism, and the future of folk psychology. *Philosophical Perspectives, 4*, 499–533.

Rorty, R. (1979). *Philosophy and the mirror of nature*. Princeton: Princeton University Press.

Ryle, G. (1948). *The concept of mind*. London: Hutchinson.

Searle, J. (1992). *The rediscovery of the mind*. Cambridge, MA: MIT Press.

Seibt, J. (1990). *Properties as processes: A synoptic study of Wilfrid Sellars's nominalism*. Atascadero, CA: Ridgeview.

Sellars, W. (1963). *Science, perception and reality*. London: Routledge & Kegan Paul.

Sellars, W. (1968). *Science and metaphysics*. London: Routledge & Kegan Paul.

Sellars, W. (1971). Science, sense impressions, and sense: A reply to Cornman. *The Review of Metaphysics, 23*, 391–447.

Sellars, W. (1981). Foundations for a metaphysics of pure process. *The Monist, 64*, 3–90.

Stich, S. (1983). *From folk psychology to cognitive science*. Cambridge, MA: MIT Press.

Tuomela, R. (1977). *Human action and its explanation*. Dordrecht: Reidel.

Tuomela, R. (1984). *A theory of social action*. Dordrecht: Reidel.

Tuomela, R. (1985). *Science, action, and reality*. Dordrecht & Boston: Reidel.

Tuomela, R. (1988). The myth of the given and realism. *Erkenntnis, 29*, 181–200.

Tuomela, R. (1991). On radical conceptual revolutions in social science. *Journal for General Philosophy of Science, 22*, 303–320.

Tuomela, R. (1994). *The importance of us: A philosophical study of basic social notions*. Stanford, CA: Stanford University Press.

Wittgenstein, L. (1953). *Philosophical investigations*. Oxford: Blackwell.

11 In Search of the Science of Consciousness

Antti Revonsuo
University of Turku, Finland

INTRODUCTION

The return of consciousness to the psychological sciences seems, at the moment, inevitable. After a lengthy exile from the world of research and carrying a reputation of "bad science," consciousness finally strikes back. The number of books and articles published on this subject during the last few years is amazing, and the amount of interest and attention philosophers and cognitive scientists have devoted to consciousness is steadily growing.

However, there is not yet any genuine paradigm of consciousness research: There are no universally accepted core assumptions on which a general theory of consciousness could be built. There certainly are models or preliminary theories of consciousness (Baars, 1988; Crick & Koch, 1990; Damasio, 1989, 1990; Dennett, 1978, 1991b; Dennett & Kinsbourne, 1992; Edelman, 1989; Flanagan, 1991, 1992; Jackendoff, 1987; Schacter, 1990), but is there any common core to all this theorizing? Which empirical findings are relevant to the theory of consciousness? What are the philosophical commitments of such a theory, or rather what should they or should not be? Current theories of consciousness, it appears, have no common core because they stem from diverse philosophical foundations and because they are confined to taking into account only a portion of all the relevant empirical findings. I suggest that we ought to look at consciousness simultaneously as a philosophical and an empirical problem. Only by combining these views into a coherent whole can we establish a true paradigm of consciousness research.

The first and most profound question we must face—and on which the different theories do not agree—is this: What basically *are* conscious phenomena? How do they fit in with the scientific picture of the rest of the world? More specifically, can the existence of consciousness be explained (or even illuminated) by referring to, say, neurobiological phenomena, or should we rather, in our explanations, refer to the abstract functional or computational levels? Is consciousness ultimately a neurobiological, functional, computational, or some completely different kind of property? I call this question that inquires into the fundamental nature of consciousness the *ontological problem*. If the science of consciousness aims to be a progressive branch of research, it must have a clear answer to this question. Otherwise, the different theories are based on different and mutually incompatible principles and cannot be expected to give compatible predictions, explanations, or hypotheses about empirical phenomena. It seems that the more philosophically minded theorists have a specific ontology as their starting point, and they try to fit consciousness into this philosophically motivated world view, often ignorant of most or all of the empirical findings related to conscious phenomena. Empirical cognitive scientists, by contrast, do not explicitly confront the ontological problem at all but rather answer it only implicitly. Despite this, they are eager to talk about consciousness in the light of the empirical data, forgetting that the philosophical groundwork remains to be done.

The second question is, or at least it can be formulated as, purely empirical. Among neuroscientists, this problem is often called the *binding problem* (Crick & Koch, 1990; Damasio, 1989). It refers to the discrepancy between our experiences of reality and the mechanisms in the brain supposedly underlying those experiences in some way. The world "as-perceived" (Velmans, 1990), that is, the experienced world, is a coherent whole in which perceptions of color, form, space, movement, sound, touch, and smell are combined to create an experience of one world in which three-dimensional colored well-formed objects and localizable sounds surround the perceiver. By contrast, neuroanatomy and neurophysiology suggest that all normal perception involves processing in anatomically fragmented and functionally different cortical regions and that there seems to be no unitary system in which this shattered puzzle is being put together. How is it possible to explain perceptual unity at the phenomenological level, when all that we can point to at the level of brain processes is neurophysiological divergence? How can we perceive an object through vision, hearing, touch, smell, taste, or understand its linguistic name, and experience it as one and the same object? We must have a genuine theory to explain why the world seems to us united and why the contents of consciousness, the experienced world, seem to us a determinate and coherent whole.

Therefore, the empirical question to be asked is: By virtue of which subsystem or processing mode of the brain is the experienced world bound together to form phenomenological coherency? The proposed models of consciousness by no means agree on how the binding problem should be resolved. Some of them fail to acknowledge the whole issue. Others suggest that there must be some sort of central system in the brain into which all the rivers of sensation flow, and it is there that the coherent representation of the world is constructed. In some models, this kind of a "consciousness-centered" solution is rejected, and consciousness is thought to arise in a fragmentary way.

Because the currently available models of consciousness do not agree on even what the basic structure of such a model should be, all of them simply cannot be on the right track. Thus, we need to find out which approaches to consciousness are fruitful and which are not. Only a consensus about the basic principles of the science of consciousness can guide future empirical research. Otherwise, the matter will once again be doomed to scientific oblivion. In this chapter I briefly review the most important models of consciousness that have been introduced recently. My purpose is to find out the answers they give to the key questions mentioned previously: the ontological problem and the binding problem. I then argue that if we simultaneously take into account philosophical arguments and a lot of empirical evidence, we can dramatically reduce the number of possible answers to these two problems. Although no single model alone can take us very far, a suitable combination of certain philosophical ideas with a mass of empirical findings can, I believe, show us the way.

The five models of consciousness I consider are the following: the multiple drafts model (Dennett, 1991b; Dennett & Kinsbourne, 1992), the computational model (Jackendoff, 1987), the cognitive model (Baars, 1988), the neuropsychological model (Schacter, 1990; Schacter, McAndrews, & Moscovitch, 1988) and the neurobiological model (Crick & Koch, 1990; Damasio, 1989, 1990). (For a review, see Revonsuo, 1993a.) In addition, I consider the eliminative alternative as a possible answer to the ontological question. It is not a model of consciousness but rather a set of arguments for giving up the modeling of consciousness.

ELIMINATION OF CONSCIOUSNESS

The *eliminative alternative* is roughly an attempt to deny the reality of any "problems of consciousness." Consciousness is an element of common sense thinking and folk psychology, but there can be no useful scientific characterization of it. We have no precise theories, definitions, or operationalizations of consciousness because such things are simply impossible. A scientific theory of consciousness is, according to eliminativists, as inconceivable as a theory of phlogiston, gnomes, or geosentric cosmology. We cannot have such theories, because in reality there exists nothing of which they would be theories. The history of science teaches that people have believed in many things that never really existed; they were making false ontological presuppositions. The concept of consciousness is obscure and heterogenous. Any science trying to incorporate consciousness into its ontology will be making a fatal mistake; such a science plainly cannot survive for long because of its fundamental incorrectness. Thus, the eliminativist answer to the ontological question, "What is the nature of consciousness?" is, "Nothing, because no such thing really exists!" (For eliminativist views, see Allport, 1988; Churchland, 1988; Sloman, 1991a, 1991b; Stanovich, 1991; Wilkes, 1984, 1988.)

We can argue against elimination along the following lines. First, from the present nonexistence of precise theories and definitions of consciousness, it does not follow that such things are forever impossible. The research has just begun, and an account of conscious phenomena seems to be needed in many fields (e.g., Baars, 1988; Schacter & al., 1988; Young & de Haan, 1990). Consciousness seems all but futile for science, and its characterizations can be scientifically refined, as I try to show in this chapter. So, elimination is not a viable alternative at the moment: We should give the science of consciousness at least a fair chance to have a try before condemning it. In addition, elimination offers no solutions, no theoretical answer to the binding problem that remains completely ignored and no less enigmatic. If believing in consciousness means advocating a thoroughly mistaken ontology, the least the eliminativists should do is to provide an explanation of why we seem to enjoy an experienced reality. Phlogiston theories of combustion were overthrown by showing that burning in fact involves oxygen, not that combustion does not exist; geocentric theories were given up by showing how the seeming geocentrity could result from a heliocentric, or rather from a centerless cosmology, not by claiming that the earth does not exist. The binding of the experiential world, accordingly, must be explained either by showing how true binding occurs in the brain or by showing how a fragmented reality in the brain can result in a coherent and determinate phenomenology in consciousness. It surely cannot be solved by insisting that scientifically, conscious experiences are not worth talking about (Revonsuo, 1993b).

THE MULTIPLE DRAFTS MODEL

The ideas behind the *multiple drafts model* of consciousness have been developed by the philosopher Daniel C. Dennett in numerous writings (Dennett, 1978, 1982, 1987, 1990), but it is only presently that together with Marcel Kinsbourne, he introduced it as a whole (Dennett, 1991b; Dennett & Kinsbourne, 1992; see also Dennett, chapters 2 & 6, this volume). What, then, are Dennett's answers concerning the ontology and the binding of consciousness? Dennett (1987) separates psychology into two fundamentally dissimilar explanatory realms: *subpersonal cognitive psychology* and *intentional system theory*. The latter is "a sort of holistic logical behaviorism" and "deals with the prediction and explanation from belief-desire profiles of the actions of whole systems," and "the *subject* of all the intentional attributions is the whole system rather than any of its parts" (Dennett, 1987, p. 58). Now it is of principal relevance to understand what the ontology of the entities is about which the intentional system theory talks. The concept of *belief* is a splendid example. In Dennett's view, beliefs cannot be discovered inside the brains of organisms. Beliefs are *interpretations* attributed to intentional systems by external observers, but in reality there are no such things. Concepts of intentional system theory are *observer-relative*, that is, they can be utilized by the researcher in any way that he finds serviceable for characterizing, explaining, or predicting the system's behavior. No doubts should arise whether or not the attributions are correct, because the attributed states, like beliefs, do not exist out there, independently of external observers. An intentional interpretation of a system can be good or bad, practical or futile, but never true or false.

Dennett writes that "all there is to being a true believer is being a system whose behavior is reliably predictable via the intentional strategy" (Dennett, 1987, p. 29). "There is no magic moment in the transition from a simple thermostat to a system that really has an internal representation of the world around it" (Dennett, 1987, p. 32). And because there is no particular "correct" interpretation of the contents of beliefs and other mental states, it follows that "no further fact could settle what the intentional system in question *really* believed" (Dennett, 1987, p. 40). "There could be two different systems of belief attribution to an individual which differed *substantially* in what they attributed . . . and yet where no deeper fact of the matter could establish that one was a description of the individual's *real* beliefs and the other not. . . . The choice of a pattern would indeed be up to the observer, a matter to be decided on idiosyncratic pragmatic grounds" (Dennett, 1991a, p. 49). Consequently, Dennett supposes that it is "a mistaken conviction that our own beliefs and other mental states must have determinate content" (Dennett, 1987, p. 42).

The concepts of subpersonal cognitive psychology radically differ from those of intentional system theory. Whereas the latter concepts designate observer-relative "calculation bound entities or logical constructs", the former concepts refer to "posited theoretical entities" (Dennett, 1987, pp. 53–57). That is, the subpersonal theory describes the actual events in the brain in a functionalistic language. The final target for the subpersonal theory is to decompose the cognitive functions of the organism into more and more elementary subsystems, the operations of which can finally be constructed at the implementational level (neurophysiology). This box-model of cognition, or positing of many "stupid" homunculi to explain the behavior of the whole, must be carried out in such a manner that the homunculi do not duplicate the system-level properties they are supposed to explain (Dennett, 1978).

In Dennett's world view, the ontological status of consciousness is crucially dependent on whether its proper place is in intentional system theory or in subpersonal cognitive psychology. It strongly seems that consciousness, for Dennett, belongs to intentional system theory, and accordingly, it shares the ontology of the other entities of that theory. He says that the subject of the access of personal consciousness is the person and not any of the person's parts (Dennett, 1978, p. 150) and that a subpersonal model "evades" the question of personal consciousness (Dennett, 1978, p. 154). Moreover, Dennett (1987, p. 42) asserts that it is "a mistaken conviction" that the contents of our mental states are determinate. The same also concerns the contents of consciousness: His theory (1991b, p. 113) "avoids the tempting mistake of supposing that there must be a single narrative that is canonical—that is the *actual* stream of consciousness of the subject."

In this view, contents of consciousness are ascribed to—not found in—intentional systems capable of realizing the right sorts of organism-environment interaction (For Dennett's reply to this criticism, see Revonsuo, in press). The functional description of such systems can ultimately be made at an algorithmic program level, and thus any entity capable of instantiating that program has a mind. "An appropriately programmed computer . . . literally has a mind, whatever its material instantiation, organic or inorganic" (Dennett, 1990, p. 50).

All in all, the ontological status of consciousness in the multiple drafts model is that of an observer-relative label that can be ascribed to a whole organism or system but not to any of its parts. Furthermore, in this framework, the ontology of mind in general is based on the abstract program-level that enables the system to behave successfully and intelligently in its environment.

How, then, does Dennett handle the binding problem? He strictly denies that there is any "center" in the brain "where it all comes together." He argues that nothing in functional neuroanatomy suggests that such a center exists and that postulating the property of consciousness to some subpersonal component defies the general strategy of subpersonal theories: It assumes a system-level property to a subsystem and thus only duplicates the problem at that level (Dennett, 1978, 1991b; Dennett & Kinsbourne, 1992).

Dennett's "solutions" to the ontological and the binding problems are, to say the least, eccentric. He seems to propose (1991b) that we should conduct some sort of self-imposed brainwashing to get rid of our confused intuitions according to which we really possess determinate contents of consciousness and according to which there really is such a thing as a phenomenological "experienced world." He states (1991b) that "*There seems to be phenomenology*. . . . But it does not follow from this undeniable, universally attested fact that *there really is* phenomenology. This is the crux" (p. 366). Also that "there is no way to isolate the properties *presented* in consciousness from the brain's multiple reactions to its discriminations, because there is no such additional presentation process" (p. 393). Perhaps this is true, but at least we should be granted an explanation of why there seems to be phenomenology; why it inescapably appears that an experienced world is presented for us in consciousness. Dennett utterly fails to give such an explanation. He thinks that we should try to ignore or change our phenomenological intuitions by force rather than by seeking a theory to resolve the binding problem. In other words, he does not take the binding problem seriously—he only questions if, in the first place, there is anything there which is being bound, and he tries to show that at least there is nothing in the brain that could accomplish such binding. In this sense, Dennett's theory fares no better than the eliminativist alternative: Neither of them provides genuine answers; they only deny the question (for a more detailed critique of Dennett's philosophy of consciousness, see Revonsuo, 1993b).

In addition, Dennett's view implies the empirically questionable claim that the line between unconscious and conscious processes cannot be drawn inside the brain. For Dennett, it is a category mistake to try to find out which brain events are the basis of conscious phenomena and which are blind and unconscious: consciousness can be ascribed only at the level of a whole person or intentional system. However, there are anatomical structures in the central nervous system (e.g., the cerebellum) that clearly function as nonconsciously as, say, the thyroid gland, and there are conceptual divisions in cognitive and physiological psychology which closely resemble the division between conscious and unconscious processes (automatic vs. controlled processing; modular vs. central systems; see e.g. Baars, 1988; Fodor, 1983; Näätänen, 1990; Shiffrin & Schneider, 1977). Moreover, empirically based theories of conscious phenomena (Baars, 1988; Crick & Koch, 1990; Damasio, 1989; Schacter, 1990) are indeed trying to find the borderlines between the conscious and the nonconscious inside the brain.

The multiple drafts model hardly provides any basis for the science of consciousness, because it distorts the meaning of consciousness; consciousness is an observer-relative label *ascribed* to whole intentional systems but not anything *real* going on in the brain of an intentional system. It denies the binding problem and apparently also the worth or observer-independent existence of subjective, experienced phenomenology. Furthermore, Dennett believes that organisms and robots can have minds in virtue of the programs they implement, if the program just enables "the right sorts of environment-organism sequences of interaction to occur" (Dennett, 1990, p. 50). This "computer program of the mind" can be defined disregarding entirely the actual material composition of the brain. Later I consider arguments to the effect that such a computationalist ontology, often popular within defenders of artifical intelligence (AI), could not serve as the basis for a science of consciousness.

THE COMPUTATIONAL MODEL

The *computational model* of consciousness is argued for by Ray Jackendoff (1987). He thinks that there is such a thing as "the computational mind" and that phenomenology is somehow "caused by/supported by/projected from" certain representational levels in the computational mind (p. 298). Thus, the nature of the mechanism by which phenomenology arises from computations remains mysterious. Jackendoff (1987, p. 18) comments that "it is completely unclear to me how computations, no matter how complex or abstract, can add up to an experience." Another philosophical worry springs from his "hypothesis of the nonefficacy of consciousness" (pp. 24–25), according to which consciousness in itself cannot have any causal effects on computations; causality works solely from the computational mind to the phenomenological mind. This means, as he observes, that consciousness is no good for anything: It becomes a useless epiphenomenon, a genuine ghost in the cognitive machinery.

Jackendoff tries to solve the binding problem by suggesting that there is a "selection function" at each representational level that selects one representation and puts it into short-term memory. People are not aware of all these selected representations, only of those projected from appropriate levels in each modality of awareness. Linguistic awareness, for example, is projected from phonological structure, musical awareness from what Jackendoff called the "musical surface."

Although Jackendoff sincerely tries to solve the binding problem, his ontological solution is simply disastrous, which undermines his good intentions. It is utterly mysterious how computations could be identified with or give rise to conscious experiences. Consequently, Jackendoff postulates an obscure one-way projection or causation that extracts consciousness from computations but that simultaneously renders it an ineffectual will-o'-the-wisp. Such an epiphenomenon hardly improves an organism's chances of survival or production of offspring. Jackendoff's theory is, then, antibiological and antievolutionary. It can hardly be a good starting point for future theories of consciousness.

WHAT IS THE ONTOLOGY OF COMPUTATION?

For many biologically minded readers, it might be a bit unclear what exactly it is that philosophers and AI-advocates mean when they talk about computations as the basis of mind. What are computations, ontologically speaking? I believe that this question has not been asked often enough and even more rarely clearly answered. I consider the possibility, pointed out by John Searle (1990, 1992, and chapter 4, this volume) that computations are ontologically such things that no satisfactory theory of consciousness could rest on such an ontology. The two previous theories of consciousness—the multiple drafts model and the computational model—are, I am afraid, the best that we can get if we presuppose a computationalist ontology. The former theory represents computationalism combined with a radical externalization of consciousness and the rejection of the binding problem; the latter is an attempt to solve the binding problem within a computationalist framework. Why do these solutions not work?

We can shed some light on what computations are by comparing them with natural, biological, chemical, or physical properties. Computations are defined at an abstract level, independently of the physical systems in which those computations might be realized. The notion of a universal Turing machine is the theoretical idea of a computer that can compute any well-defined algorithmic computations. Modern general-purpose digital computers basically incarnate this idea. Thus, to identify a computation, we need to know nothing of the materials of the machines in which the computation is carried out: If those machines satisfy a certain abstract functional description, they all can be said to compute the same function, no matter whether they are made out of grains of sand, silicon-chips, solar systems, or banana peels. In principle (though hardly in practice) the universal Turing machine can be built out of any arbitrary bits of matter (for an illustrious example, see Maudlin, 1989).

All of this means that if we can, by some wizardry, find out which computations or functions of the brain are those that underlie our conscious experiences, then we can program them into any arbitrary universal Turing machine, and while the computations are being executed, the machine enjoys a similar consciousness as we do. Now, how could some computations differ from others so radically as to be able to produce consciousness? This is the core question, and the alternative answers to it are manifested in Dennett's and Jackendoff's theories. Computations cannot *in themselves* have such properties which could explain the relevant difference. The only escapes are in the externalization, or the mystical projection, of consciousness. For Dennett, the content of a (computational) mental state is not to be found in the state itself (as a neurophysiological state) but needs interpretation from the outside intentional stance. Dennett and Kinsbourne (1992, pp. 235–239) say that the creation of conscious experience is a continuous process in which "microtakings" interact. A microtaking "cannot just be inscribed in the brain in isolation"; "it has to have its consequences," for example, for "guiding action." "Becoming a conscious experience does not clearly endow an event with potencies it previously lacked . . . No one stream is necessarily conscious by its very nature. And that *is* our theory of consciousness" (Dennett & Kinsbourne, 1992, pp. 238–239). It follows that computations or programs can be (treated as) conscious, provided only that they enable the implementing system to perform appropriate interaction with the environment. Hence, according to Dennett and Kinsbourne, consciousness is not a subjective mental phenomenon but something logically dependent on behavior. By contrast, for Jackendoff, some computations in themselves really are (or at least they produce) conscious states, independent of the external behavior or third-person interpretations. However, because it is an absolute riddle how this is possible, an eerie mechanism must be assumed that, in the end, leaves consciousness overhanging somewhere above computations, crippled and futile.

In sharp contrast to this, it is, at least in principle, possible to explain how different physical, chemical, and biological systems can have different kinds of properties at different levels of natural organization, independent of how we decide to interpret or treat those systems. We can safely make the reasonable assumption that there is only one hierarchically arranged world and that consciousness is a property manifested at some level in this hierarcy. How, in general, does this hierarchy with its different levels work? If computations cannot add up to conscious states, how could material particles fare any better? It seems that there is some sort of *part–whole dependence* between different levels of organization: The property of liquidity can only be realized in systems composed of something like molecules but not in systems composed of photons or of planets. Similarly, hydrogen atoms, photosynthesis, DNA-molecules, mitosis, hurricanes, and supernova explosions all require their own specific micro level conditions to occur; they certainly cannot be had in arbitrary pieces of matter, like computations can. Philosophers have developed a technical term *supervenience* to describe, among other things, these kinds of part–whole relationships in nature, and they suggest that each higher-order property or level requires *a minimal supervenience basis* at the micro level on which the macro level depends (Kim, 1984, 1990; Kincaid, 1988; Rowlands, 1990, 1991). For example, the minimal basis for something to be an instatiation of a hydrogen atom is that at the subvenient level, it has one proton and one electron. However, the number of neutrons might be one, two, or none.

Now if consciousness is a part of this natural order reaching from gluons to galaxies, it also needs its minimal supervenience basis, that is, it can only be realized in certain sorts of materials or neurobiological systems, not in abstract computations. If we cannot have liquidity, metabolism, photosynthesis, or thunderstorms in computations, why on earth should we expect to have consciousness in them? John Searle (1990, 1992, chapter 4, this volume) has recently argued that it is a serious mistake to think of the brain as a "biological computer" with "computations" going on somewhere inside. Computation is standardly defined in terms of "symbol manipulation" and "syntax," but there are no such things in nature itself: Nothing out there, independent of us, can be a symbol in itself. We can never find symbols or computations in nature, but we can assign computational interpretations to any systems we choose. Thus, computation ontologically implies the notion of a symbol, and a symbol entails an external observer who treats a physical thing in nature as a symbol. Brains compute no more than planets solve differential equations about their orbits, or flowers want to turn to light. Consciousness somehow naturally supervenes on the physical properties of the brain, like other natural properties in their own organizational levels do. Programs and computations, however, do not exist in themselves. They are observer-relative and reside only in the eye of the beholder; they are no part of the physical and causal levels. Between quarks and quasars, there is nowhere a natural organizational level called "the program level", but there are such levels as molecules, cells, organs, mental phenomena, ecosystems, and stars. Our tendency to see programs or computations in the brain stems from our ignorance of the real causal, probably neurobiological, level.

It might be objected that *connectionist AI* escapes the criticisms presented here. Perhaps a connectionist, nonsymbolical model of a conscious process could be the same as the real thing, after all. Unfortunately, as far as pure computations are concerned, no relevant difference exists between connectionist and traditional AI: Any function that can be computed with a connectionist machine can also be computed with a traditional machine (and vice versa). So we cannot say that conscious processes are functions that could only be computed in parallel machines. If conscious processes are essentially computational, then they can be programmed to serial as well as to connectionist machines.

There seems to be a perilous misconception plaguing the kind of connectionist research that aims at modeling brain function. For example, Sejnowski, Koch, and Churchland (1990, p. 8) declare that "one of the major research objectives of computational neuroscience is to discover the algorithms used in the brain." They think that the brain cannot be regarded as a digital computer because the "computational solutions evolved by nature" are constrained by the actual architecture of the brain system. However, it can be considered a "naturally evolved" connectionist computer. Consequently, Churchland and Sejnowski (1992, p. 66) believe that "finding out what is computed by the cerebellum will teach us a lot about the nature of the tissue and how it works." Contrast this view of computational neuroscience with the way computational methods are used in other branches of natural science. For example, Rohrlich (1990) examines a computational model of the atomic level that depicts a nickel pin contacting the plane surface of a piece of gold and assumes 5000 atoms in dynamical interaction. Another computational model mentioned by him describes the evolution of a spiral galaxy. Now, we may ask, why is it that the major research objectives of (computational) physics or astronomy is not, as far as we know, to *discover* the algorithms used in the atoms or to find out what is computed by the stars and galaxies? The answer seems simple in the extreme: Because atoms and galaxies do not *use* algorithms, they do not compute anything. There are no algorithms in natural phenomena to be discovered by us. By contrast, we must find out what the causal interactions going on in those systems actually are, and then we may hope to describe those interactions with the help of an algorithmic computational model.

Even Churchland and Sejnowski (1992) at one point admit the interest-relative nature of computations and computational systems: " We count something as a computer because, and only when, its inputs and outputs can usefully and systematically be interpreted as representing the ordered pairs of some function that interests us. . . . This means that delimiting the class of computers is not a sheerly empirical matter, and hence that 'computer' is not a natural kind, in the way that, for example, 'electron' or 'protein' or 'mammal' is a natural kind" (p. 62). Now, if computations and computational systems are not natural kinds, neither are they natural phenomena that could be "discovered" or "found out" by us. This exactly is Searle's (1992, p. 212) point when he argues that computational cognitive science cannot ever be a natural science, because natural science tries to discover features intrinsic to the natural world. Computation is not such an intrinsic feature that could ever be discovered: It has to be assigned relative to observers. Thus, although Churchland and Sejnowski appear to accept Searle's premise, according to which "computer" (and "computation") names no natural property or process but an observer-relative one, the profound consequences of Searle's argument seem to escape from them.

The point then is not that the architecture of the brain constrains the set of algorithms that can be used in it. This way of thinking grasps the problem in the reverse order. Imagine that we find the algorithms that, for example, compute the phenomenological variable of brightness, or perceived intensity of light from the physical variable of luminance, or the amount of light arriving at the retina (Grossberg & Todorovic, 1990). Probably the architecture of the brain and the nature of the problem call for highly specific algorithms to resolve this computational problem. Imagine that we have found very satisfactory solutions to this problem: The output values (representing brightness) and computed from the input values (representing luminance) almost perfectly match the relevant empirical evidence. We may say that we have discovered the algorithms that are used in the brain. Two problems arise. Any computational system has an infinite number of possible interpretations (Kukla, 1992), so the system not only represents the computation of brightness from luminance but also all other isomorphic sets of functions. Without us acting as interpreters, the mapping between input and output does not represent anything at all. Second, although the computations in the brain are constrained by the architecture, could we not, after having discovered them, implement them in digital machines as well? There certainly is no computational restriction. If all the magic is in the algorithm, the digital computer ought to duplicate the powers of the brain despite the differences in the architecture, thus reproducing, say, perceived brightness. We end up with phenomenology mysteriously arising from computations, like Jackendoff did in his theory of consciousness.

So, what is wrong? It is the idea that the computation or algorithm has a causal or explanatory role in the system instead of just describing the system formally. The architecture of the brain does not constrain the algorithms it uses or the computations it performs. Instead, the causal relationships inside the brain, between different levels of organization and inside the elements comprising one level of organization, are so complex and specific that they can be described by only very specific computational models. The formal character of any system is not in the system itself qua a material and causal system, but only in our descriptions of the system. The specific algorithms by which we are able to describe and model brain function are a symptom of the specific organizational levels of nature that, in the final analysis, are the causes and explanations of mental phenomena. William Uttal (1990) reminds cognitive theorists that mathematical or computational formal models of cognition are excellent descriptions but not explanations of the phenomena in question: "Formal models . . . can have nothing definitive to say about the physiological or logical mechanisms by means of which these processes are carried out" (p. 189). "Computational, mathematical, or neural net models can, in principle, say nothing rigorous about internal mechanical reality in a reductionistic sense . . . regardless of how perfectly they describe function and process" (p. 199). Searle (1992) returns to the same problem noting, like Uttal, that the cognitivist program falsely promises to reveal us the causal mechanisms of cognition: "The mechanisms by which brain processes produce cognition are supposed to be computational, and by specifying the programs we will have specified the causes of cognition. . . . But without a homunculus, both commercial computer and brain have only patterns, and the patterns have no causal powers in addition to those of the implementing media. So it seems there is no way cognitivism *could* give a causal account of cognition" (pp. 215–216).

In fact, most of the connectionist programs that exist today are simulations of real networks. That is, the serial computer has a formal description of a network, and it changes the activation values one by one, according to the input, ending up with the same result a real network would have; the digital computer is mimicking network behavior. Let us then assume that we build a real connectionist network and not just a simulation. Such a connectionist system is not any more so clearly independent of the implementation level (although its input-output relations could be simulated in traditional machines) but by contrast, seems to need its own "minimal supervenience basis" of units, connections, and weights that roughly is a basis also realized in the neuronal networks of the brain. Although connectionism undoubtedly is a step in the right direction, it certainly as such does not explain how conscious processes are realized in the brain. Daniel Schacter suspects that "systems depicted by PDP models are largely implicit memory machines: When a new pattern of activation partially matches an existing pattern of activation, the existing pattern is automatically strengthened and affects the response tendencies of the system, without any recollective experience or conscious awareness of a past event" (Schacter, 1989, p. 709).

Many if not most of the neural networks involved in human cognition are working totally nonconsciously (for instance the perceptual representation systems underlying implicit memory, Tulving & Schacter, 1990). Consequently, the difference between traditional and connectionist AI is not the same as the difference between nonconscious and conscious processes in the brain. To date, connectionism has not been able to illuminate very much the nature of the mechanisms underlying the specifically conscious processes in the brain. Of course, once we have an actual scientific account of the mechanisms of consciousness, there will be nothing preventing us from modeling them with computational models. But, we must remember, the computational simulation of conscious processes has exactly the same relationship to actual conscious processes as the simulation of weather phenomena has to actual winds and rains.

To summarize the discussion, computationalism is an inadequate ontological basis for theories of consciousness because:

1. Computational systems cannot have different levels of organization in the sense that physical systems can and do. Likewise, they cannot have such part–whole relations on which the different properties at different levels of organization in nature are based. Because of this, Jackendoff and others have failed in telling us how computations in themselves could add up to experiences or indeed to any natural phenomena at all, as physical systems do.
2. The notion of computation denotes nothing in nature itself but implies reference to an external interpreter: Computations are assigned, not found in nature. Thus, Dennett's view that conscious states and the whole of phenomenology are not in nature itself but are instead observer-relative interpretations made from the intentional stance is only a natural concomitant of a computationalist ontology. Identification of computations or programs in the

brain is observer-relative as is, in Dennett's theory, identification of conscious and other mental states.

3. So far, even connectionism has not illuminated the specifically conscious processes of the brain. Computational neuroscience, based on (connectionist) modeling of the brain, is misguided if it is assumed that there really is computation or vector transformation going on in the brain—independently of us—in the same sense as there are high-frequency oscillations in certain neuronal populations, for example. Computations and algorithms are our descriptions of the systems at a formal level. There is no calculation going on in any natural phenomenon (excluding mental arithmetics).

The cognitive model, the neuropsychological model, and the neurobiological model remains. The advocates of these models have made practically no explicit commitments on the ontological problem but have concentrated on the binding problem. However, Edelman (1989) clearly rejects computationalism, and Crick and Koch express some fairly critical but unphilosophical opinions about it: "We believe that the problem of consciousness can, in the long run, be solved only by explanations at the neural level. Arguments at the cognitive level are undoubtedly important, but we doubt whether they will, by themselves, ever be sufficiently compelling to explain consciousness in a convincing manner. . . . A major handicap is the pernicious influence of the paradigm of the von Neumann digital computer" (Crick & Koch, 1990, pp. 263–264).

For empirical researchers it is, presumably, difficult to recognize the deep ontological problems underlying computationalism. They might just feel some vague uneasiness in front of a rival philosophical view, unable to put their finger on what exactly is wrong with it. Searle's (1990) critique of computationalism might be what they have been missing: "Computational models are useful in studying the brain, but nobody really believes that the computational model is a substitute for a scientific account of how the system actually works, and they should stop pretending otherwise" (p. 637).

A REALIST ONTOLOGY FOR CONSCIOUSNESS

Because the computationalist answer to the ontological question as well as the multiple drafts model and the computational model have now been examined and found to be wanting, we must find a different course to take. Do we still have hope for adequate answers to the ontological and the binding problems? I believe we do. I suggest that the existence of consciousness should be illuminated and explained by referring to a lower level of real and natural organization, say, neurobiology. This means that phenomenology—the world-as-experienced—is itself a level of organization in nature, a natural phenomenon. This natural phenomenon is brought about by and realized in the neurobiology of the brain. This does not necessarily mean that psychology should or could be reduced to neuroscience; even biology itself has not been and probably never will be reduced to chemistry (Kincaid, 1988, 1990). Supervenience and the unity of science are consistent with nonreducibility (Kim, 1990). It also does not mean that no machines could ever have a mind: Mental properties, like other properties of physical systems, have minimal supervenience bases, and within these bases, they are multiply realizable. The exact minimal supervenience basis of any natural phenomenon is an empirical question, no matter whether the natural phenomenon be hydrogen, black hole, or consciousness. This ontological approach has many advantages:

1. It places consciousness among other natural phenomena in the hierarchy of the universe.
2. It treats consciousness as something real and determinate and not dependent of external observers.
3. Consciousness is seen as causally efficacious in behavior and selected for in evolution.
4. It admits the existence of phenomenology.

The science of consciousness, I suggest, is looking for something like the following:

1. There must be a system that has a "phenomenological level of organization"; a system that constitutes phenomenal experience, the experienced world. The functioning of this level is the necessary and sufficient condition of conscious experience. At the macro level of organization, the system is characterized by referring to the world-as-experienced by the organism. At the micro level of organization, the system is described in terms of the neurobiological mechanisms that are in charge of bringing about the world-as-experienced.
2. There must be an explanation of how the phenomenological level of organization can be or seems to be (for the subject) a multimodal, coherent, and united whole of color, form, distance, movement, sound, smell, touch, and so forth. That is, there must be a theory that resolves the binding problem

and explains how this system can have access to such a wide variety of perceptual and somatosensory information. This theory is a crucial link between the micro and macro levels of organization.

3. There must be a theory to explain why this experiential world is a biologically useful model of the world. It should explain how the organism can adapt its behavior, through any output mode, to the real physical world that in itself, of course, is totally beyond any experiences. The experiential model of the world (which is, no doubt, species specific) enables the organism to increase the likelihood of its survival and production of offspring. If for any reason any information that is normally available to conscious experience becomes unavailable (e.g., neurological disruption), then it also ceases to be available for adaptive behavior (this is manifested in neuropsychological patients with implicit knowledge, Lahav, 1993; Schacter et al.,1988; Young, chapter 8, this volume).

The remaining three approaches can be seen as sharing an ontology that is more or less like that stated previously and as trying to find answers to these questions. I believe that the cornerstones of the future science of consciousness lie somewhere here, and in the latter part of my chapter I try to sketch an approach that combines the ontological with the empirical.

THE NEUROPSYCHOLOGICAL MODEL

The *neuropsychological model* is based on the curious phenomenon of implicit knowledge in neuropsychological syndromes. The pattern of functional dissociation is generally something like the following: After brain injury, a patient is found to be severely impaired in performing certain cognitive operations, say, remembering recent events, recognizing faces, or comprehending words. The patient usually complains about these defects and is painfully aware of his handicaps. Also the patient's overt behavior and test performance reflect his cognitive inabilities. However, if the same cognitive function is tested indirectly or implicitly (e.g., as revealed through galvanic skin responses, evoked brain potentials, priming effects, or interference), the patient might perform significantly better, sometimes even normally (on these tests). It seems that such a patient has implicit, unconscious knowledge of the stimuli although he cannot explicitly or consciously access the same information (for reviews, see Bruyer, 1991; Schacter et al., 1988; Schacter, 1990; Young, chapter 8, this volume; Young & de Haan, 1990).

Implicit knowledge is information that is revealed in performance without the subject's phenomenal awareness of possessing any such information. The existence of implicit knowledge in a wide variety of neuropsychological syndromes is a test for cognitive neuropsychology: What does this dissociation tell us about the normal organization of cognition and consciousness? Schacter et al. (1988) stressed the universality of the pattern of dissociation, and they thought it is useful to search for a common explanation for this pattern. They put forward a descriptive model that is based on the following assumptions:

1. Conscious or explicit experiences of perceiving, knowing, and remembering all depend in some way on the functioning of a common mechanism. In other words, conscious awareness of a particular stimulus requires the involvement of a mechanism that is different from those mechanisms that process the various attributes of the stimulus.
2. This conscious mechanism, or *conscious awareness system* (CAS) (Schacter, 1990), must be sharply distinguished from modular systems. CAS normally accepts input from and interacts with a variety of processors or modules that handle specific types of information.
3. Conscious experience of various types of information depends on intact connections between the conscious awareness system and individual modules. In cases of neuropsychological impairment, specific modules can be disconnected from the conscious mechanism.

This model postulates the existence of a common system for all kinds of conscious processes. It is theoretically useful to refer to such a system if we want to understand the observed patterns of dissociation in neuropsychological patients. Thus, the conscious awareness system is, in the neuropsychological model, the way to approach the binding problem. Despite the disunity and isolation of sensory processing going on at the modular input level, it is rational to postulate the existence of a single, unifying system at the level of conscious experience. Although merely assuming a system such as CAS is still far from a genuine or final explanation of experiential unity, it is encouraging that empirical evidence points to the existence of a system that may, at least in principle, be responsible for the coherency of the world-as-experienced.

THE COGNITIVE MODEL

Bernard J. Baars (1988, and chapter 7, this volume) takes as a starting point the contrasts between unconscious and conscious psychological processes. The key idea here is to contrast two psychological processes, the only dissimilarity between which is that one of them is conscious and the other not. On the basis of such analyses, Baars proposed that conscious and unconscious events are realized in different kinds of processing architectures in the nervous system.

The basic architecture behind unconscious processing is that of specialist processors that according to Baars, are functionally unified or modular. The specialized, unconscious processors are very efficient and rapid, they make relatively few mistakes, and an assembly of such processors may operate in parallel without interfering with each other. Specialized processors are relatively isolated and independent, and they can maneuver only certain kinds of information. This characterization of specialist processors comes quite close to what have been called "modules" in cognitive neuropsychology (Ellis & Young, 1988; Fodor, 1983).

The architecture behind conscious processes is entirely different. In studies of attention, it has been shown that there seems to be a central attention system with limited capacity. Experiments on selective attention have shown that subjects can be conscious of only one coherent stream of events at any one time. Also, it has been established that similar cognitive tasks compete with each other, probably because they demand the same processing resources, and even very dissimilar tasks interfere with each other if they are both executed consciously. In general, according to Baars, conscious processes are computationally inefficient, slow, prone to errors, and interfering with each other. They can have a great range of different contents at different moments of time, but at any one moment there is only one conscious process going on that must be internally consistent. The conscious system thus has limited capacity, and it operates serially.

Conscious operations are distributed globally in the central nervous system. Given that the same information must be available for the entire system at one flash, it follows that there can be only one truly global message at any one time. There is only one whole system, and if all its elements are to receive the same message, only one message at a time can be broadcast. This is Baars' interpretation of the seriality of conscious processes. The system in which information can become globally distributed and consciously experienced is called *global workspace*. It is the communication network that unites all the unconscious processors. Conscious processing is a specific manner of processing accomplished in the global workspace.

Baars suggests that the neural substrate of the global workspace is the "extended reticular-thalamic activation system" (ERTAS). This system is known to have excellent connections to virtually every input and output channel in the brain, and it is also involved in the regulation of global states of awareness such as coma or sleep. In comparison, the specialist processors reside in the cerebral cortex. The whole picture that emerges is that cortical specialist processors compete or cooperate with each other to get their own message through to the global workspace of ERTAS. A message that gets through has to be, among other things, globally broadcast, coherent, and informative. Such messages then become the contents of the stream of consciousness.

Thus, the *cognitive model* also assumes that one common system, or to be accurate, that one specific processing mode of a common system, underlies all conscious phenomena. Its solution is clearly analogous to the neuropsychological model: Both postulate a large number of unconscious "knowledge modules" (Schacter, 1990) or "specialist processors" (Baars, 1988), but conscious phenomena in both models depend on a common system that is functioning above the modular level. The cognitive model, however, is presenting more accurate hypotheses concerning the necessary and sufficient conditions as well as the neurobiological substrates of conscious processes.

THE NEUROBIOLOGICAL MODEL

Whereas the neuropsychological and the cognitive models try to identify, mostly at the macro level of description, the system in charge of consciousness, the *neurobiological model* tends to operate at the borderline between the macro and micro levels of organization. That is, the former models attempt to characterize the relationships that the "box" of consciousness has with the other unconscious "boxes" in the information processing models of mind. In contrast, the neurobiological model aims at probing the insides of the postulated conscious box or system, thus uncovering the actual causal mechanisms that operate at the lower level of organization, bringing about the macro level of conscious experience. The model I call here the neurobiological model has actually been presented by independent researchers, especially Damasio (1989, 1990) and Crick and Koch (1990) (but to some extent, also Calvin, 1990; Edelman, 1989; Llinás & Paré, 1991; Posner & Rothbart, 1991). However, these models have so much in common in their basic principles that I believe it is perhaps not unjustified to treat them under the same title.

The starting point here is the total fragmentation of the brain processes underlying experiences, and the task is to find out how this neuronal puzzle can ever form coherent experiences: "Any entity or event that we normally perceive through multiple sensory modalities must engage geographically separate modality structures of the central nervous system" (Damasio, 1989, p. 28). Also "The information about a single object is distributed about the brain" (Crick & Koch, 1990, p. 274).

In 1984, Francis Crick published an important paper in which he outlines some possible neural mechanisms for perceptual unity. He argues that we must look for neural mechanisms manifesting very short-term transient alterations that occur during the act of visual perception. He gives the hypothetized synapses capable of this feat the name Malsburg synapses. In Crick's view, there must be transient vertical assemblies of cells that unite the general concept (e.g., "face") with parts of a perceived face and further with the details of those parts. The members of such vertical assemblies of cells must be able to fire synchronously.

Damasio (1989, 1990) denies, on neuroanatomical grounds, the traditional view according to which an integrated representation of an object or scene is formed somewhere in a unified neuron group at a single anatomical site. The neural activity and the representations that experience can access reside in a fragmented fashion and in geographically separate areas located in modal sensory cortices. The integration of these fragments into experiences is based on the synchronous, time-locked activation of the anatomically separate sites. When we perceive an object as a coherent entity, separate regions at the early sensory cortices (and in the motor and somatic cortices as well) are activated. The activation patterns correspond to the different features such as color, form, and texture of the object. The point is that these feature-encoding patterns become synchronously activated. According to Damasio, the key to the coherent and simultaneous activation of these patterns is coded at the so called convergence zones. They are functional regions that have feedback connections to the primary regions of activation. Local convergence zones code the activation patterns within one modality so that the features define, for example, the shape and the color of an object. Nonlocal convergence zones encode more complex combinations of features, also across modalities. Thus, a local zone might activate the color, shape, and texture of a banana, whereas the nonlocal zone will, in addition, activate the smell, taste, feel, and maybe the name of the object. Thus, the convergence zones are seen as stores of combinatorial codes responsible for the binding of features into coherent entities in consciousness and recall. The synchrony takes place in primary cortical areas, it is paced from other higher-order cortical areas (convergence zones), and the activity can be modulated by projections from the thalamus and subcortical areas. According to Damasio, the success of this time-locked activation is dependent on attention, which is defined as a critical level of activity in each of the activated regions below which consciousness cannot occur.

Crick and Koch (1990) suggest that binding (especially in the visual modality) involves an attentional mechanism that synchronizes the neuronal spikes in frequency-locked oscillations between 35 and 75 Hz. It has been shown theoretically that a set of oscillating neurons quickly become frequency- and phase-locked if there is a centralized feedback unit rather than diffuse connections to the pacemaker. It is unclear where in the brain the oscillations are generated and paced or how the relevant set of distinct neurons is selected. Crick and Koch believe that the selection could be achieved through competition of different alternative interpretations of the same sensory information. We become aware of the best interpretation, and the neurons coding its features become frequency locked. The rest of cortical activity remains unconscious.

Edelman (1989, p. 72) claims that so called "reentrant" signaling or processing between "mapped regions" allows "the various responses to aspects of the stimulus world to remain segregated and distributed among multiple brain areas and still constitute a unified representation. Reentrant integration obviates the need for a higher-level command center." In addition, Edelman (1989) emphasizes that consciousness is biologically useful: It helps the animal to satisfy its adaptive needs and to guide its behavior to reach particular goals.

Posner and Rothbart (1991) argue that any mechanism capable of integration between different systems must share information from many sources and it should have a close relationship to motor systems. Such integration, they believe, is at the heart of the problem of consciousness: "According to a number of views, this co-ordination does not require a part of the brain that holds within its anatomy the information which is to be integrated. It is possible instead to argue that parts of the brain are responsible for acting upon codes that are anatomically quite separate" (Posner & Rothbart, 1991, p. 103).

In sum, the neurobiological model denies that representations would in some concrete way be bound or glued together in a mysterious part of the brain (essentially the same view as Dennett's). However, it takes the binding problem seriously by proposing that the splinters of experiences become bound together through synchronous, time-locked activation or frequency-locked oscillations or reentrant signaling. Thus, we have here a common basis for all kinds of conscious phenomena and some specific hypotheses concerning the neural mechanisms underlying the integration manifest at the level of conscious experience. This elucidates the mystery of how a shattered neural reality could ever form the basis of the unified phenomenology we in fact enjoy.

IS CONSCIOUSNESS DISTRIBUTED OR CENTRALIZED?

At first glance, the remaining models of consciousness disagree on one fundamental question. The neuropsychological model and the cognitive model suggest that consciousness is realized in some sort of centralized and unifying system, whereas the neurobiological model denies this and proposes that consciousness is a property of distributed neural activations. Both views are based on empirical findings, albeit of different kinds. Can this disagreement be resolved? If we are able to find, despite the manifest discrepancy, a common core of these views, the prospects for a unified theory of conscious phenomena are enhanced. If, however, no such resolution is to be found, the science of consciousness appears to be doomed for the time being. Let us examine more closely how serious this dilemma actually is.

The implicit view embedded in the neuropsychological and the cognitive models is that from the unconscious modules or specialist processors, psychological information somehow flows to the common consciousness system. Damasio, however, explicitly denies the image of psychological contents flowing from unconscious to conscious systems in the brain: "No representations of reality as we experience it are ever transferred in the system; that is, no concrete contents and no psychological information move about in the system" (Damasio, 1989, p. 57).

In the cognitive model, the problem is to find the principles that govern the selection of information from the unconscious to the conscious. Baars (1988) claims that neuroscientists mostly work with the complex, parallel, and fast unconscious parts of the brain, whereas psychologists have been interested in the serial and slow mode of functioning. He suggests that we should try to look for a global workspace system in the brain, because such a system would explain the gap between the two modes of processing. This system should be connected with conscious functions such as wakefulness, focal attention, and orienting responses. On the input side, many systems should have access to it, and different inputs should be able to cooperate or compete when accessing the global workspace. On the output side, it ought to be able to distribute the activity and information to many other parts of the brain—practically anywhere.

Baars (1988) and Baars and Newman (chapter 9, this volume) propose that this kind of global neural workspace in the brain consists of the reticular formation of the brain stem and midbrain, the outer shell of the thalamus, and the set of neurons projecting upward diffusely from the thalamus to the cerebral cortex. He calls this system "the extended reticular-thalamic activating system" (ERTAS). "We can therefore suggest that the ERTAS underlies the 'global broadcasting' function of consciousness, while a selected perceptual "processor" in the cortex supplies the particular *contents* of consciousness to be broadcast. . . . These conscious contents, in turn, when they are broadcast, can trigger motor, memory, and associative activities. There is independent evidence that cortical activity *by itself* does not become conscious. . . . We would suggest that any cortical activity must trigger ERTAS support in a circulating flow of information before it can be broadcast globally and become conscious" (Baars, 1988, p. 126). Proponents of the neurobiological model clearly do not support this view: "Where in the brain are the neural correlates of consciousness? One of the traditional answers—that consciousness depends on the reticular activating system in the midbrain—is misleading. . . . It is more likely that the operations corresponding to consciousness occur mainly (though not exclusively) in the neocortex and probably also in the paleocortex" (Crick & Koch, 1990, p. 265).

Nonetheless, the models seem to agree that the content of a conscious state resides in the cortex, but in the cognitive model, this information content must be "broadcast" across the system, whereas Damasio strictly denied such transfer of representations. Does the neurobiological model agree with the cognitive model that cortical activity does not by itself become conscious? Evidently it does, because "attention"—whatever it is—is needed in the neurobiological model to boost the level of activity.

The notion of attention is exceptionally vague in the neurobiological model. Both Damasio (1989) and Crick and Koch (1990) treat it as a "spotlight" that metaphorically, "lights up" just the right cortical contents (by linking their frequencies?) at the right moment. However, attention is precisely the most crucial mechanism, the operation of which is imperative to understand if we want to know why some contents become conscious and others remain in the dark. It is an enigmatic process that operates in the twilight zone between the neural piecemeal representations and the mental, conscious experiences. Furthermore, the neurobiological model gives no account of the limited capacity of consciousness. It does not prevent every single representation from being conscious at the same time—granted their activity levels are boosted in the same time-window.

How, then, is attention understood in the neurobiological model? Damasio (1989, p. 49) wrote that "attention depends on numerous factors and mechanisms," which is of course true, but not very illuminating. Nevertheless, he emphasized that the state of the perceiver and the context of the process also play important roles in determining the levels of activations and that "the reticular activating system, the reticular complex of the thalamus, and the limbic system mediate such roles under partial control of the cerebral cortex."

It appears that we can, finally, settle for some agreement between the disparate theories. My interpretation is that both theories postulate the interaction of the ERTAS system with the cortical representations as an essential requirement for conscious contents to occur. It is not clear whether the information content should be thought of as moving in the system, as Baars suggest, but perhaps such a supposition is not indispensable. Anyway, the global workspace or some such system is needed to explain the limited capacity and seriality of conscious processes.

Also the neuropsychological model (Schacter et al., 1988) is in accordance with this idea, because this model implies that conscious processes require the involvement of a mechanism different from the modular discrimination systems. Edelman's (1989) "biological theory of consciousness" is broadly consistent with such a proposal, too. The basic generation mechanism of conscious experiences, according to the theory, is the interaction of interoceptive input mediated by limbic and brain-stem circuits with exteroceptive input mediated by thalamus and cortical areas. In concert with the other models, Posner and Rothbart (1991) argue for a separate attention system as the source of coordination and activation in many brain areas.

Nonetheless, there is one crucial detail in the neuropsychological findings that seems to defy the neurobiological model. If the existence of implicit knowledge at least in some cases is the outcome of normal cortical activation, as seems to be the case concluding from the normal implicit performance, then it cannot be a matter of pure activation that is at stake. In other words, the interaction that is missing brings something qualitatively new into the process, not just more activation, because normal activation per se seems not to help the patients who possess implicit knowledge to become conscious of the content or even the existence of that information.

Summarizing the discussion, to be able to have consciousness of some specific content we necessarily need the representation of that content in the appropriate cortical module. But to have any kind of consciousness at all we require the global workspace or something similar that can somehow access and interact with these contents. If we picture consciousness as always originating from the interaction between these two systems, the controversy between consciousness-centered and fragment theories begins to be lost. The contents are grounded on fragmentary representations, but the crucial interaction comes from some sort of general-purpose system. It depends on the character of this interaction whether representations are literally transferred in the system or whether they merely undergo state changes right where they are. This question is empirical. One point must be remembered, however: The assumed interaction is by no means a homunculus "looking at" representations or a modular "audience" attending a show at the Cartesian theater. Whatever the precise nature of the critical neurobiological process is, the phenomenology that results is just experienced, not observed. Once inside phenomenology, the distinction between the observer and the observed dissolves, only experience remains.

TOWARDS THE SCIENCE OF CONSCIOUSNESS

The combination of a realist ontology (which comes quite close to what Searle calls "biological naturalism") with three converging, empirically based approaches to consciousness (the neuropsychological, the cognitive, and the neurobiological models) seems to me to be the advisable choice on our way toward the scientific understanding of consciousness. Why? As Bechtel and Richardson (1993) show, there are certain rational explanatory strategies that are particularly well suited to scientific problems that are relatively ill-defined or in which neither the criteria for a correct solution nor the means for attaining it are clear. The problem of consciousness certainly seems to fit that description. In fact, it seems that numerous explanatory tasks in the life sciences have begun from such a state of confusion. Before it is possible to develop an explanation of *how* some particular phenomenon or function is brought about, it is necessary to identify, first, *what* the phenomenon or function in question is, and second, *which system* or systems is responsible for producing that effect.

In the case of explaining mental phenomena and discovering the mechanisms behind the scenes, we may use a strategy Bechtel and Richardson (1993) call "complex localization," that is, models that are *multiply constrained.* In the case of consciousness, the organization at the psychological or phenomenological level provides top-down constraints on the model, whereas the neural mechanisms provide bottom-up constraints, limiting the number of possible mechanisms underlying the phenomenological level. It seems to me that it is not only possible but also very fruitful to use such multiple constraints from different levels of organization when we attempt to have a grasp of consciousness. The empirically based models of consciousness reviewed previously are good examples of the use of rational explanatory strategies that have been used elsewhere in the life sciences for decades, if not centuries.

As an example of this development, I examine the significance of the binding problem. I think that merely asking this question guides research in the right direction, because the question implies that we take consciousness seriously and that we can try to find multiply constrained models of consciousness. When we approach a scientific problem as ill defined as consciousness is, the first task is to identify the phenomenon to be explained. To see perceptual binding as an empirical problem to be resolved serves the function of implicitly identifying consciousness as a very specific kind of phenomenon. When the researchers decide to take the binding problem seriously, it means that they (implicitly or explicitly) accept that there is, for the subject, a unified world of experience. How that experienced world can be brought about by the neurophysiological processes in the brain is the big question to be explained. That is, the structure of experience, the way consciousness manifests itself for us (as subjects), is taken for granted: It is the phenomenon that strikes us as odd in the context of neuroanatomy and neurophysiology.

As we know from the philosophy of science (e.g., Laudan, 1977), a problem is identified as a problem in science only in a certain context of inquiry. Taking the binding problem seriously implies taking into account the first-person's point of view of the world. Incidentally, I think that this is why Dennett has had great difficulties in taking the binding problem seriously. For him, taking the first-person's subjective point of view into account is, presumably, to abandon the "objective, materialistic, third-person world of the physical sciences" (Dennett, 1987, p. 5) on which Dennett's philosophy of mind is erected. Consequently, he says that there are "incautious" formulations of the binding problem in current neuroscientific research that presuppose that there is some process in the brain that puts all the various discriminations into registration with each other (Dennett, 1991b, pp. 257–258). The "incautiousness" of these formulations, I am afraid, stems from the fact that the perspective of the first person is almost built in into the binding problem. If none of us had ever experienced a coherent perceptual world surrounding ourselves, the binding problem would never have occurred to us. Were we actually living our personal lives in the "third-person world of the physical sciences," the binding problem would never have arisen. Because Dennett firmly adheres to the third-person world, he did not even attempt to solve the binding problem, because the whole problem apparently cannot be made intelligible in the terms he accepts.

At the level of psychological organization, interest was first drawn to perceptual binding by the experiments of Treisman and her colleagues (Treisman & Gelade, 1980; Treisman & Schmidt, 1982). They showed that searching visual target items that were defined by a conjunction of features (say, one vertical red bar among 100 multicolored vertical and horizontal ones) took much more time than just searching for a single feature (one vertical bar among 100 horizontal ones). In fact, searching for the conjunction required attentive scanning of about 50–70 milliseconds per stimulus, whereas the single feature just pops out immediately regardless of the number of distractors. Thus, forming a percept in which color and form are bound together takes the brain both time and attention. What happens if we test the limits of visual feature integration? When complex visual stimuli were flashed for the subjects, it was found that the subjects started to perceive illusory conjunctions of properties. That is, a red *H* and a green *T* were perceived by half of the subjects as a green *H* and a red *T*. Thus, the brain did not have enough time to finish the binding process reliably, and so it resulted in illusory combinations of properties. Note that the illusory conjuctions never existed anywhere outside the subjects' world of experience: The first person's point of view was already crawling in at this point.

At the neural level of organization, it was discovered that odor-specific information is organized as patterns of neuronal activity in the gamma frequency range (20–80 Hz), and Freeman & van Dijk (1987) reported similar oscillatory activity in the monkey visual cortex. The oscillations have been found to emerge selectively in separate sites in response to specific stimuli. Where two separate recording sites have similar response preferences, the gamma oscillations between those separate sites show increased correlation (Gray & Singer, 1989). Gray and his colleagues (Gray, König, Engel, & Singer, 1989) showed that these gamma correlations have something to do with the binding of the different visual properties into a coherent image. They recorded gamma activity from two separate sites 7 mm apart in the cat visual cortex. When two short light bars were moved in opposite directions, there was practically no correlation between the oscillations. When they were moved in the same direction, a slight correlation was observed. But when a single, long light bar was moved across the two receptive fields, the correlation was significantly increased. Thus, the gamma oscillations seem to reflect the global integrity of the stimulus.

These empirical findings are considered as highly relevant for theories of consciousness (Damasio, 1990; Crick & Koch, 1990). The coherent oscillations between multiple cortical sites can form an almost infinite number of different transient states in the brain. This combinatory power makes it intelligible how it is possible that the world we experience is so rich in phenomenological distinctions and details. Every phenomenological distinction, we assume, must have a corresponding distinction at the level of neurophysiological states. Now we can sketch a neurophysiological mechanism that could at least in principle be what we have been looking for. Consequently, trying to resolve the binding problem takes us a long way from merely drawing boxes labeled "consciousness" into our flowcharts of the mind–brain. Here we are finally starting to look for the mechanisms that in fact underlie the world-as-experienced. In effect, we are asking, "How does the brain do it? How could the multimodal, coherent world of experience, even in principle, be brought about by the combinatory mechanisms between neurons and neuron assemblies?" This means that we believe there are illuminating and isomorphic part–whole relationships obtaining between the phenomenological and the neural level of description and organization.

At the moment, what are the prospects for further research on consciousness? If the synchronous 40 Hz oscillations have anything to do with consciousness, we ought to find out more about them. For example, the neuroscientist Rodolfo Llinás (Llinás, 1993; Llinás & Paré, 1991) has discovered, by using magnetoencephalographic recordings in awake humans, that there are continuous 40 Hz oscillations over the entire cortical mantle. Moreover, this spontaneous activity becomes synchronized when auditory stimuli are presented. Interestingly, comparison of the phases between different cortical regions reveal that there is a phase shift of about 12 milliseconds between the rostral and the caudal pole of the brain. It seems that the same gamma wave is sweeping the brain from front to back, thus creating an organized macroscopic stimulus-related event in the brain. Llinás hypothesizes that cortico-thalamo-cortical pathways play an essential role in the generation and organization of these oscillations. This is because the cortical layer IV neurons are capable of creating spontaneously 40 Hz oscillations, and the reciprocal projections from the cortex back to the reticular thalamic nucleus would then result in rebound oscillation in the thalamic neurons (which are also capable of intrinsic oscillation). Thus, these cortico-thalamo-cortical pathways could then lead to resonant oscillation at the observed 40 Hz. According to Llinás' hypothesis, the reticular thalamic cells are responsible for synchronizing the 40 Hz oscillations in remote thalamic and cortical areas.

Exciting as these new findings and hypotheses are, they must be regarded as preliminary. However, here we have some fairly precise claims that can be empirically tested. This means that a natural-science approach to consciousness is finally getting off the ground. One remarkable point here is that, by studying cerebral rhytmic activities in the gamma frequency range, the neuroscientists are approaching something like the concept of global workspace, although the evidence on which Bernard Baars originally based his theory stems from completely different sources. Consider Bressler's (1990) conclusions concerning the gamma waves: "The concept that emerges from these studies is that information is integrated by a cooperative network of interconnected neurons and is manifested as spatially modulated patterns of synchronized gamma oscillation. . . . Synchronization of gamma oscillations may come to be seen as a general cortical phenomenon, integrating activity among widely separate areas of cortex. . . . It may turn out that the gamma oscillation, serving a role as cortical information carrier, is also coördinated in global patterns throughout the entire neocortex" (Bressler, 1990, pp. 161–162).

On the other hand, in the neurophysiological interpretation of the global workspace theory, Baars and Newman (chapter 9, this volume), end up with the same structures of the brain as the basis of consciousness as does, for example, Llinás in his neurobiological theory of consciousness. It is enticing to ask whether the synhcronous events that scan our brain with the observed 12 millisecond phases (that is, about 80 times per second) are nothing but the global workspace in action. "If we assume that the phase shift observed in these preliminary studies is related to the presence of simultaneity waves which scan our brain at 80 Hz, we can conclude that consciousness is not a continuous event. Rather, it is determined by the simultaneity of activity in the thalamocortical system. . . . If consciousness is a product of thalamo-cortical activity, it is the dialogue between the thalamus and the cortex that creates subjectivity" (Llinás & Paré, 1990, pp. 531–532).

If there is a promising convergence taking place with respect to the neurobiological and the cognitive models, the neuropsychological model seems to be rapidly catching up with them. In a recent paper Ran Lahav (1993) examines what neuropsychology can tell us about consciousness, and he concludes that explicit (conscious) information plays roles in the behavior of the organism entirely different from implicit, nonconscious information. The latter can influence only a very limited range of behavioral responses, its effects tend to be isolated from the global behavior of the organism, and they amount only to fixed, simple input-output relationships. In contrast, conscious information can evoke a wide range of different kinds of behavioral responses, and such responses are integrated into the organism's overall activity.

Lahav writes that "Conscious experience expresses information available for an entire spectrum of global, integrated, and flexible (non-automatic) behaviors. . . . It constitutes a central *junction of information,* one in which information from different sources and modalities are integrated to produce a unified and coherent body of behaviors" (Lahav, 1993, p. 79). In addition, he observes that there is little if any continuity between explicit and implicit information: We either have a global and flexible response to a stimulus, or we have an isolated and automatic response. The information cannot occur in a state that would be something between these two extremes: Conscious and nonconscious events really appear to express two entirely distinct types of processes in the organism's brain. The "central junction of information" revealed to us by the behavior of neuropsychological patients fits very well together with the global workspace theory, although even in this case, the evidence is coming from two entirely independent sources.

The inevitable outcome appears to be that no matter which empirical approach we take with regard to consciousness, we end up with more or less similar hypotheses concerning the nature of the underlying system. The most recommendable approach, therefore, is to use multiple, converging constraints from different levels of investigation and domains of empirical inquiry and to encourage interaction between separate levels and domains. In this way, I believe, we may finally be able to force consciousness into a corner and develop fruitful interfaces between the diverse theories of consciousness that so far have been isolated from each other—each single theory restricted to taking into account only some narrow aspect of consciousness.

CONCLUSIONS

We have now wandered through the different alternatives of the ontology and the binding of consciousness. If a true paradigm of consciousness research ever emerges, I guess that there will be only one satisfactory combination of solutions to be found to the ontological problem and the binding problem. The eliminative alternative simply denies—on dubious grounds—the binding problem and offers no theory at all instead. The multiple drafts model also ignores, or perhaps tries to suppress, the dilemma that arises from the discrepancy between phenomenology and neural reality. In addition, it treats the concept of consciousness in a way that is not only counterintuitive but also empirically rather questionable. It seems that the multiple drafts model and the computational model share a computationalist ontology. Such an ontology has no place in the natural world, as Searle (1992) persuadingly argues. If consciousness is a natural phenomenon, as I think it is, we cannot explain (or even much illuminate) it by referring to nonnatural, observer-relative phenomena such as computations.

The most appealing solution to the ontological problem, I suggest, is that we see conscious phenomena as residing at one level in the hierarchy of other natural organizational levels of nature. The same principles that govern other natural organizational levels also ought to apply to consciousness. Thus, properties in general can be realized only in rather specific systems: for something to be an atomic nucleus, a gas, a self-replicating molecule, a living cell, photosynthesis, a conscious phenomenon, a star, or a quasar, it means that this something has a specific microstructure organized in a certain way, and only microstructures sufficiently similar to each other can ever function as the basis of such phenomena. It is an empirical question, of course, what exactly the "sufficiently similar" means in each case: We need to know the minimal supervenience bases of each phenomenon. Computations, as we noted, have no minimal supervenience bases and consequently, are no part of the natural organization of things, but, by contrast, are abstract and observer-relative. Consciousness, however, is part of the natural organization of things and is by no means abstract or observer relative. Thus, the computational model was rejected because of its highly debatable ontology.

Finally, we were left with three models that were all consistent with the realist solution of the ontological problem, seeing consciousness as a real phenomenon at the organizational level of the neurobiology of human and animal (and possibly some other) brains. We resolved their apparent conflict concerning the binding problem by proposing that, in fact, all the models speak of approximately the same thing. First, there is one general mechanism: (a) in the cognitive model, the global workspace or ERTAS; (b) in the neuropsychological model, the conscious awareness system; (c) in the neurobiological model, the spotlight of attention realized as synchronous activation or frequency-locked oscillations. This is the system that takes care of the coherent and time-locked activation.

Second, there is one fragmented system—the sensory areas, modules, or specialist processors of the cortex—that provide the contents of conscious states. One unresolved question is whether psychological information should be thought of as moving in the system, or whether we should think of the representations of contents undergoing state changes right where they are.

Be that as it may, it does not change the point that we have found a common core of models of consciousness that combines a satisfactory ontology with serious attempts to resolve the binding problem. In addition, we noted the remarkable convergence that can even at this early stage be seen between the completely different and independent empirical approaches to consciousness. In the search of the science of consciousness, the best sign of being on the right track is that it is leading us somewhere. And even if the chosen path would not lead us anywhere, we know that considering multiple constraints from different levels of organization and investigation is a rational strategy that has been successfully (and unsuccessfully) used in numerous parallel scientific problems in the past (Bechtel & Richardson, 1993). So, if there will never be any science of consciousness, this kind of multidisciplinary exploration probably is a reasonable method by which to find out that important fact. No matter how strongly philosophers would like to settle the question of consciousness, they will not be able to do it without considering all the empirically relevant data. It might turn out that empirical research will soon leave behind those models of consciousness that do not offer convincing solutions to the ontological and binding problems.

REFERENCES

Allport, A. (1988). What concept of consciousness? In A. J. Marcel & E. Bisiach (Eds.), *Consciousness in contemporary science* (pp. 159–182). New York: Oxford University Press.

Baars, B. J. (1988). *A Cognitive theory of consciousness*. New York: Cambridge University Press.

Bechtel, W., & Richardson, R. C. (1993). *Discovering complexity: Decomposition and localization as strategies in scientific research*. Princeton, NJ: Princeton University Press.

Bressler, S. L. (1990). The gamma wave: A cortical information carrier? *Trends in Neurosciences, 13(5)*, 161–162.

Bruyer, R. (1991). Covert face recognition in prosopagnosia: A review. *Brain & Cognition, 15*, 223–235.

Calvin, W. (1990). *The cerebral symphony: Seashore reflections on the structure of consciousness*. New York: Bantam.

Churchland, P. S. (1988) Reduction and the neurobiological basis of consciousness. In A. J. Marcel & E. Bisiach (Eds.), *Consciousness in contemporary science* (pp. 273–304). New York: Oxford University Press.

Churchland, P. S., & Sejnowski, T. J. (1992). *The computational brain*. Cambridge, MA: MIT Press.

Crick, F. (1984). Function of the thalamic reticular complex: The searchlight hypothesis. *Proceedings of the National Academy of Sciences, 81*, 4586–4590.

Crick, F., & Koch, C. (1990). Towards a neurobiological theory of consciousness. *Seminars in the Neurosciences, 2*, 263–275.

Damasio, A. R. (1989). Time-locked multiregional retroactivation: A systems-level proposal for the neural substrates of recall and recognition. *Cognition, 33,* 25–62.

Damasio, A. R. (1990). Synchronous activation in multiple cortical regions: A mechanism for recall. *Seminars in the Neurosciences, 2,* 287–296.

Dennett, D. C. (1978). Toward a cognitive theory of consciousness. In D. C. Dennett (Ed.), *Brainstorms* (pp.149–173). Brighton: Harvester.

Dennett, D. C. (1982). How to study human consciousness empirically or nothing comes to mind. *Synthese, 53,* 159–180.

Dennett, D. C. (1987). *The intentional stance*. Cambridge, MA: MIT Press.

Dennett, D. C. (1990). The myth of original intentionality. In K. A. Mohyeldin Said, W. H. Newton-Smith, R. Viale, & K. V. Wilkes (Eds.), *Modelling the mind* (pp. 43–62). Oxford: Clarendon Press.

Dennett, D. C. (1991a). Real patterns. *Journal of Philosophy, 88,* 27–51.

Dennett, D. C. (1991b). *Consciousness explained*. Boston: Little, Brown.

Dennett, D. C., & Kinsbourne, M. (1992). Time and the observer: The where and when of consciousness in the brain. *Behavioral and Brain Sciences, 15,* 183–247.

Edelman, G. (1989). *The remembered present: A biological theory of consciousness*. New York: Basic Books.

Ellis, A. W., & Young, A. W. (1988). *Human cognitive neuropsychology*. Hove & London: Lawrence Erlbaum Associaties.

Flanagan, O. (1991). Consciousness. In O. Flanagan (Ed.), *The science of the mind* (2nd ed, pp. 307–366). Cambridge, MA: MIT Press.

Flanagan, O. (1992). *Consciousness reconsidered*. Cambridge, MA: The MIT Press.

Fodor, J. A. (1983). *The modularity of mind*. Cambridge, MA: MIT Press.

Freeman, W. J., & van Dijk, B. W. (1987). Spatial patterns of visual cortical fast EEG during conditioned reflex in a monkey. *Brain Research, 422,* 267–276.

Gray, C. M., König, P., Engel, A. K., & Singer, W. (1989). Oscillatory responses in cat visual cortex exhibit inter-columnar sychronization which reflects global stimulus properties. *Nature, 338,* 334–337.

Gray, C. M., & Singer, W. (1989). Stimulus-specific neuronal oscillations in orientation columns of cat visual cortex. *Proceedings of the National Academy of Sciences (USA), 86,* 1698–1702.

Grossberg, S., & Todorovic, D. (1990). Solving the brightness-from-luminance problem: A neural architecture for invariant brightness perception. In S. J. Hanson & C. R. Olson (Eds.), *Connectionist modeling and brain function: The developing interface* (pp. 393–420). Cambridge, MA: MIT Press.

Jackendoff, R. (1987). *Consciousness and the computational mind*. Cambridge, MA: MIT Press.

Kim, J. (1984). Concepts of supervenience. *Philosophy and Phenomenological Research, XLV,* 153–176.

Kim, J. (1990). Supervenience as a philosophical concept. *Metaphilosophy, 21,* 1–27.

Kincaid, H. (1988). Supervenience and explanation. *Synthese, 77,* 251–281.

Kincaid, H. (1990). Molecular biology and the unity of science. *Philosophy of Science, 57,* 575–593.

Kukla, R. (1992). Cognitive models and representation. *The British Journal for the Philosophy of Science, 43,* 219–232.

Lahav, R. (1993). What neuropsychology tells us about consciousness. *Philosophy of Science, 60,* 67–85.

Laudan, L. (1977). *Progress and its problems*. London: Routledge & Kegan Paul.

Llinás, R. R. (1993). *Consciousness and brain rhythms*. Paper presented at the Brain & Cognition Symposium, University of Helsinki.

Llinás, R. R., & Paré, D. (1991). Of dreaming and wakefulness. *Neuroscience, 44,* 521–535.

Maudlin, T. (1989). Computation and consciousness. *The Journal of Philosophy, 86,* 407–432.

Näätänen, R. (1990). The role of attention in auditory information processing as revealed by event-related potentials and other brain measures of cognitive function. *Behavioral & Brain Sciences, 13,* 201–288.

Posner, M. I., & Rothbart, M. K. (1991). Attentional mechanisms and conscious experience. In A. D. Milner & M. D. Rugg (Eds.), *Neuropsychology of consciousness* (pp. 91–111). London: Academic Press.

Revonsuo, A. (1993a). Cognitive models of consciousness. In M. Kamppinen (Ed.), *Consciousness, cognitive schemata, and relativism* (pp. 27–130). Dordrecht: Kluwer Academic Publishers.

Revonsuo, A. (1993b). Is there a ghost in the cognitive machinery? *Philosophical Psychology, 6(4),* 387–405.

Revonsuo, A. (in press). The multiple drafts model and the ontology of consciousness. *Behavioral and Brain Sciences.*

Rohrlich, F. (1990). Computer simulation in the physical sciences. In A. Fine, M. Forbes, & L. Wessels (Eds.), *PSA 1990, Vol. 2* (pp. 507–518). East Lansing, MI: Philosophy of Science Association.

Rowlands, M. (1990). Anomalism, supervenience, and Davidson on content-individuation. *Philosophia, 20,* 295–310.

Rowlands, M. (1991). Towards a reasonable version of methodological solipsism. *Mind & Language, 6,* 39–57.

Schacter, D. L. (1989). Memory. In M. I. Posner (Ed.), *Foundations of cognitive science* (pp. 683–725). Cambridge, MA: MIT Press.

Schacter, D. L. (1990). Toward a cognitive neuropsychology of awareness: Implicit knowledge and anosagnosia. *Journal of Clinical and Experimental Neuropsychology, 12(1),* 155–178.

Schacter, D. L., McAndrews, M. P., & Moscovitch, M. (1988). Access to consciousness: Dissociations between implicit and explicit knowledge in neuropsychological syndromes. In L. Weiskrantz (Ed.), *Thought without language* (pp. 242–278). Oxford: Oxford University Press.

Searle, J. R. (1990). Who is computing with the brain? Author's response. *Behavioral and Brain Sciences, 13(4),* 632–640.

Searle, J. R. (1992). *The rediscovery of the mind.* Cambridge, MA: The MIT Press.

Sejnowski, T. J., Koch, C., & Churchland, P. S. (1990). Computational neuroscience. In S. J. Hanson & C. R. Olson (Eds.), *Connectionist modeling and brain function: The developing interface* (pp. 5–35). Cambridge, MA: MIT Press.

Shiffrin, R. M., & Schneider, W. (1977). Controlled and automatic human information processing: II. Perceptual learning, automatic attending and a general theory. *Psychological Review, 84,* 127–190.

Sloman, A. (1991a). Developing concepts of consciousness. *Behavioral and Brain Sciences, 14,* 694–695.

Sloman, A. (1991b). *Why consciousness is not worth talking about?* Paper presented at The Second International Colloquium on Cognitive Science, San Sebastian, Spain.

Stanovich, K. E. (1991). Damn! There goes that ghost again! *Behavioral and Brain Sciences, 14,* 696–698.

Treisman, A. M., & Gelade, G. (1980). A feature integration theory of attention. *Cognitive Psychology, 12,* 97–136.

Treisman, A. M., & Schmidt, H. (1982). Illusory conjuctions in the perception of objects. *Cognitive Psychology, 14,* 107–141.

Tulving, E., & Schacter, D. L. (1990) Priming and human memory systems. *Science, 247,* 301–306.

Uttal, W. (1990). On some two-way barriers between models and mechanisms. *Perception & Psychophysics, 48,* 188–203.

Velmans, M. (1990).Consciousness, brain, and the physical world. *Philosophical Psychology, 3,* 77–99.

Wilkes, K. V. (1984). Is consciousness important? *British Journal for the Philosophy of Science, 35,* 223–243.

Wilkes, K. V. (1988). --, yìshì, duh, um, and consciousness. In A. J. Marcel & E. Bisiach (Eds.), *Consciousness in contemporary science* (pp. 16–41). New York: Oxford University Press.

Young, A. W., & de Haan, E. H. F. (1990). Impairments of visual awareness. *Mind & Language, 5(1),* 29–48.

Author Index

Q

R

S

T

U

V

W

Y

Z

Subject Index

D

E

Z